Separation and Purification by Crystallization

ACS SYMPOSIUM SERIES **667**

Separation and Purification by Crystallization

Gregory D. Botsaris, EDITOR
Tufts University

Ken Toyokura, EDITOR
Waseda University

Developed from a symposium sponsored by the
International Chemical Congress of Pacific Basin Societies at the
1995 International Chemical Congress of Pacific Basin Societies

American Chemical Society, Washington, DC

Library of Congress Cataloging-in-Publication Data

Separation and purification by crystallization / Gregory D. Botsaris, editor, Ken Toyokura, editor.

p.   cm.—(ACS symposium series, ISSN 0097–6156; 667)

"Developed from a symposium sponsored by the International Chemical Congress of Pacific Basin Societies at the 1995 International Chemical Congress of Pacific Basin Societies, Honolulu, Hawaii, December 17–22, 1995."

Includes bibliographical references and indexes.

ISBN 0–8412–3513–9

1. Crystallization—Congresses. 2. Separation (Technology)—Congresses. 3. Chemicals—Purification—Congresses.

I. Botsaris, G. D. II. Toyokura, Ken, 1933–   . III. International Chemical Congress of Pacific Basin Societies (1995: Honolulu, Hawaii) IV. Series.

QD548.S45   1997
660′.2842—dc21   97–11250
                                                                      CIP

This book is printed on acid-free, recycled paper.

PRINTED IN THE UNITED STATES OF AMERICA

Advisory Board

ACS Symposium Series

Foreword

THE ACS SYMPOSIUM SERIES was first published in 1974 to provide a mechanism for publishing symposia quickly in book form. The purpose of this series is to publish comprehensive books developed from symposia, which are usually "snapshots in time" of the current research being done on a topic, plus some review material on the topic. For this reason, it is necessary that the papers be published as quickly as possible.

Before a symposium-based book is put under contract, the proposed table of contents is reviewed for appropriateness to the topic and for comprehensiveness of the collection. Some papers are excluded at this point, and others are added to round out the scope of the volume. In addition, a draft of each paper is peer-reviewed prior to final acceptance or rejection. This anonymous review process is supervised by the organizer(s) of the symposium, who become the editor(s) of the book. The authors then revise their papers according to the recommendations of both the reviewers and the editors, prepare camera-ready copy, and submit the final papers to the editors, who check that all necessary revisions have been made.

As a rule, only original research papers and original review papers are included in the volumes. Verbatim reproductions of previously published papers are not accepted.

ACS BOOKS DEPARTMENT

Contents

STUDIES RELATED TO INDUSTRIAL CRYSTALLIZERS
AND PROCESSES

CRYSTALLIZATION OF PARTICULAR ORGANIC COMPOUNDS

CRYSTALLIZATION OF PARTICULAR INORGANIC COMPOUNDS

INDEXES

Preface

A MAJORITY OF CHEMICAL INDUSTRIAL PROCESSES involve, at least at some stage, the separation of a compound by crystallization or its purification by recrystallization. Considerable research has been done in recent years in the crystallization field, and important findings have led to a better understanding of the crystallization phenomenon and to considerable improvements in the practice. These changes, however, reveal new levels of complexity in the crystallization process. At the same time, technological advances necessitate the use of crystallization in new processes, such as biochemical processes, and in the production of new compounds, such as biomolecules. For these reasons, the level of interest in crystallization has been continually increasing.

To satisfy this increased interest, international meetings dedicated to crystallization are organized, such as the one sponsored every three years by the European Federation of Chemical Engineering. In addition, both the American Chemical Society and the American Institute of Chemical Engineers include crystallization symposia in their national and international meetings.

This volume was developed from a crystallization symposium presented at the 1995 International Chemical Congress of Pacific Basin Societies (Pacifichem '95) in Honolulu, Hawaii, December 17–22, 1995. The symposium included 50 contributions, presented in oral and poster sessions, from nine countries.

This volume contains 23 chapters organized in five sections. More than a third of the chapters deal with the crystallization of polymorphic compounds that can crystallize in more than one form and the crystallization of chiral compounds. This emphasis is a reflection of the current interest in these two fields. The other four areas are basic studies, studies related to industrial crystallizers and processes, and two areas involving crystallization of particular organic and inorganic compounds. A different organization of the book would have shown that 10 of the papers deal with aspects of crystallization of proteins, amino acids, and pharmaceuticals, a sign of the increased activity in this area of crystallization.

Acknowledgments

This volume is the result of the efforts of many individuals. We thank the symposium participants and contributors, the authors of the chapters, the members of the organizing committee of Pacifichem '95 for giving us the opportunity and the help to organize and conduct the symposium, and the staff of ACS Books for their assistance and patience.

GREGORY D. BOTSARIS
Department of Chemical Engineering
Tufts University
Medford, MA 02155

KEN TOYOKURA
Department of Applied Chemistry
Waseda University
3–4–1 Okubo, Shinjuku-ku
Tokyo 169
Japan

January 22, 1997

BASIC STUDIES

Chapter 1

Crystal Morphology Predictive Techniques To Characterize Crystal Habit: Application to Aspirin ($C_9H_8O_4$)

Paul Meenan

DuPont, Experimental Station, Building 304, Room A205,
Wilmington, DE 19880–0304

Simulations of the crystal morphology of aspirin ($C_9H_8O_4$) have been made based upon geometrical and lattice energy models. Studies of the crystal chemistry have allowed the morphological importance to be ranked, based upon hydrogen bonding, surface charge and steric effects. The effect of morphology on particle performance is cursorily reviewed. A comparison between the predictions and growth morphologies has been made and an outline of current molecular modeling studies aimed at understanding the crystal habit-solvent dependence is given.

In any crystallizing system, there are many factors that can significantly influence particle shape and properties. Particles of the same material may be produced with different crystal structures (polymorphism), particle size (fines formation), shape (poor washing and filterability), and particle properties (attrition resistance, caking, flowability, dissolution rate and bio-efficacy). Hence, a key issue is particle engineering - "tailoring" particles with improved properties (e.g. shape, defects etc.) to optimize both production and materials handling. These are all issues of significant importance in the high value added batch crystallization procedures found especially within the agricultural and pharmaceutical industries, where batch to batch variations can lead to variations in the aforementioned properties, all of which have major ramifications to the manufacturer and end-user.

The crystal properties can be related to the intermolecular forces between molecules, and hence the application of solid state modeling techniques are playing an increasingly important role as a tool to aid in the visualization and explanation of complex crystallization phenomena.

This paper will focus on the application of solid state modeling techniques to one particular particle property - crystal habit. A morphological study on the analgesic aspirin ($C_9H_8O_4$) will be presented in order to assess the effects of morphology on processing, and materials handling. The crystal chemistry will be examined, in order to elucidate the influences of solvent on crystal habit and hence the ramifications to processing.

Theory

The theories behind crystal morphology modeling have been well documented by a number of authors *(1-4)* and hence it is not necessary to elaborate in great detail. Depending on the degree of accuracy required, one may perform an initial morphology calculation based on lattice geometry considerations, or a more complex calculation based on lattice energy simulations. The geometrical model developed by Bravais and Friedel *(5,6)* and later refined by Donnay and Harker *(7)*, allowed the relative growth along a specific crystal plane (hkl) to be related to unit cell information and space group symmetry. The law can be stated "Taking into account submultiples of the interplanar spacing d_{hkl} due to space group symmetry, the most important crystallographic forms will have the greatest interplanar spacings." This model can rapidly provide morphological data, however the level of accuracy of the prediction is subject to the degree of anisotropy in the intermolecular bonding within the crystal structure. The attachment energy model is based upon the Periodic Bond Chain (PBC) models developed by Hartman & Perdok *(1)*. They identified chains of bonds within the crystal structure known as Periodic Bond Chains (PBC's). The weakest bond within the PBC's is the rate determining step and governs the rate of growth along the direction of the chain. The fundamentals of slice and attachment energy have been developed from this theory. The lattice energy (E_{latt}) can be calculated by summing all the interactions between a central molecule and all the surrounding molecules. The lattice energy can also be partitioned into slice and attachment energies, where the slice energy (E_{sl}) is defined as the energy released upon the formation of a slice of thickness d_{hkl}. Upon attachment of this slice to a crystal surface a fraction of the lattice energy is released, known as the attachment energy (E_{att}). The assumption is made that the attachment energy is proportional to the growth rate and that the larger the attachment energy, the larger the growth rate and hence the less important the corresponding form within the morphology. This assumption has been shown to be valid for a number of growth theories *(4)*.

In the description of the intermolecular bonding, the Lennard-Jones 6-12 potential function *(8)* is one of the most common, consisting of an attractive and repulsive contribution to the van der Waals component of the lattice energy (V_{vdw}) as shown in Equation 1. "A" and "B" are the atom-atom parameters for describing a particular atom-atom interaction and "r" is the interatomic distance. This potential function has formed the basis of a variety of different force fields *(9-11)* that were utilized in this paper. A modified (10-12) version of this potential can also be employed *(10,11)* to describe hydrogen bonding. The 10-12 potential is very similar in construction to Equation 1 except that the attractive part is dependent on r^{10} rather than r^6.

$$V_{vdw} = -A / r^6 + B / r^{12} \qquad (1)$$

The electrostatic interaction (V_{el}) is determined using Equation 2 where "qi" and "q_j" are the charges on atoms "i" and "j", "D" is the dielectric constant and "r" the separation distance. In practice, these charges may be obtained from the literature (9), or more commonly have to be calculated using semi-empirical quantum mechanical program suites (12,13). In this study, the charges were either assigned with the particular force field, or were calculated using MOPAC (12), a semi empirical quantum chemistry code.

$$V_{el} = qi \cdot q_j / D \cdot r$$
(2)

Results & Discussion

Aspirin crystallizes in a monoclinic $P2_1/c$ space group with 4 molecules per unit cell and unit cell parameters of a=11.433Å, b=11.395Å c=11.395Å; β=95.68° (14); as shown in Figure 1. As a carboxylate system, there is a propensity to form hydrogen bonds, which in this case leads to the formation of hydrogen bonded dimer pairs. The predicted morphology based on the Bravais-Friedel-Donnay-Harker laws (BFDH) (Figure 2) reveals a somewhat tabular, roughly hexagonal shaped morphology, with dominant {100} forms and {011}, {002}, {110}, {11-1} and {10-2} forms also present. Lattice energy predictions were then obtained (15), using a variety of force fields (9-11). It is desired that a good agreement is noted between the predicted and observed lattice energies, as this indicates that the intermolecular force field chosen accurately describe the intermolecular interactions. The predicted lattice energies (Table I) are all in the range (26-34) kcalmol^{-1}; however an "experimental" lattice energy could not be located for comparison. The predicted values, however appear to be reasonable for the size and type of this molecule. Based on these lattice energy calculations, a number of morphologies were predicted and are shown in Figures 3-5. In general the attachment energy models, when compared to the geometrical BFDH model, exhibit a reduction in the {11-1} and {10-2} forms with a corresponding increase in importance of the {100} and {002} forms. This leads to an elongation of the predicted morphologies along the b crystallographic direction, when compared to that predicted from the BFDH model. The reason for this shift in morphological importance is that the geometrical model does not take into account the presence of hydrogen bonding, which will increase the growth rate and reduce the morphological importance of faces containing components of hydrogen bonding, e.g. the {11-1} and {10-2} forms, compared to those faces with lesser or no components of hydrogen bonding present, e.g. the {100} forms. A summary of the morphological importance's in the predicted morphologies is given in Table II.

Table I: Predicted lattice energies of aspirin

Actual /kcalmol^{-1}	Lifson et al (9) /kcalmol^{-1}	Momany et al (10) /kcalmol^{-1}	Scheraga et al (11) /kcalmol^{-1}
-	33.89	25.87	26.93

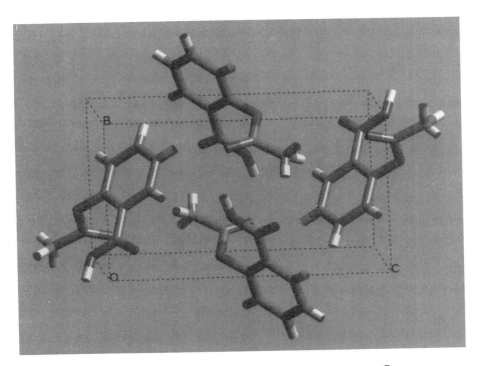

Figure 1: Unit cell of aspirin visualized using the CERIUS® molecular modelling package.

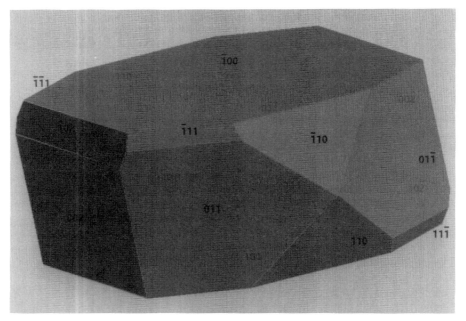

Figure 2: Predicted morphology of aspirin, based on the geometrical laws of Bravais-Friedel-Donnay-Harker (BFDH).

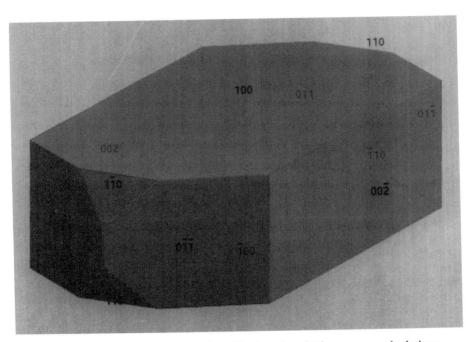

Figure 3: Predicted morphology of aspirin, based on lattice energy calculations, using the forcefield by Lifson et al (9).

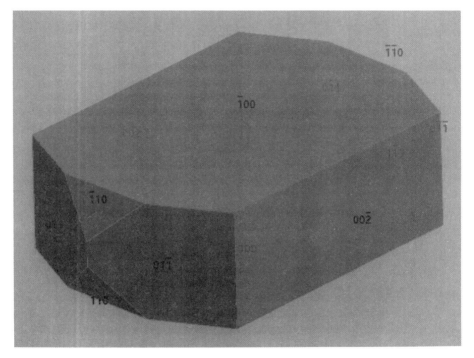

Figure 4: Predicted morphology of aspirin, based on lattice energy calculations, using the forcefield by Momany et al (10).

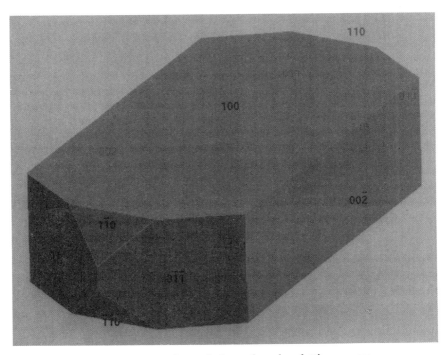

Figure 5: Predicted morphology, based on lattice energy
calculations, using the forcefield by Scheraga et al (11).

Table II: Summary of morphological importance of the different crystal forms obtained from lattice geometry and lattice energy simulations

BFDH	Lifson et al (9)	Momany et al (10)	Scheraga et al (11)
{100}	{100}	{100}	{100}
{011}	{002}	{002}	{002}
{002}	{011}	{011}	{011}
{110}	{110}	{110}	{110}
{-111}	-	-	-
{-102}	-	-	-

The surface chemistry will dictate the degree of bonding between the crystal surface and the oncoming entities, which will in turn dictate the growth rate and hence the morphological importance of the particular face. For example, a number of faces in the prediction can be examined, via "cleaving a surface from a structure. This was carried out using CERIUS® (16). The {100} surfaces (Figure 6) particularly show the dimerization of aspirin via hydrogen bonding, with weaker van de Waals forces between the dimer pairs. The hydrogen bonding displayed in Figures 6-10 has been defined as the "classic" donor-acceptor hydrogen bonding schemes defined within the distance 2Å. "Special" hydrogen bonds, e.g. C-H:::O=C, which take place over longer distances (17,18), have not been considered in this paper and consequently are not displayed. At the {100} surface, carbonyl and methyl groups predominate. The lack of hydrogen bonding projecting from the face, in conjunction with the possible steric hindrances between the methyl groups and the carbonyl groups may lead to the simulated slow growth of this face, hence the reason for the dominance of this face. The {002} forms (Figure 7) are also large, with a surface similar to the {100} forms, i.e. carbonyl and methyl groups in close proximity. Another feature of this face is the hydrogen bonding components approximately parallel to the surface, hence this bonding would not be expected to contribute to the growth rate of this face. The {011} and {110} faces (Figure 8 & 9 respectively) are very similar in nature, with greater charge density at the surface via acid and aromatic groups and also components of hydrogen bonding both parallel and perpendicular to the surface. This increased charge density and hydrogen bonding can lead to an increased surface energy and hence increased predicted growth rates. The {11-1} and {10-2} forms (e.g. Figure 10) are superficially very similar to the (011) and (110) faces and were of lower morphological importance in the BFDH model. A small increase in the importance of the {002} forms or the {011} forms would lead to the disappearance of these forms, which is indeed what happens upon a habit prediction.

Comparison Between Prediction and Experiment. Figure 11 is an illustration of a number of different growth habits obtained from different solvents (19). The wide variety of morphologies obtained may be in part due to the polarity of the solvent or due to the solubility of aspirin in the particular solvent, i.e. a supersaturation-habit dependence. The predicted forms are in reasonable agreement with the forms observed. What is readily apparent is the dominance of the {100} forms in the growth crystals obtained from the different solvents, which is in excellent agreement with predicted morphologies. The predictions are in excellent agreement with the

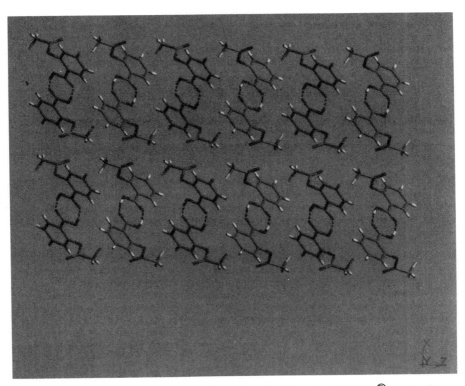

Figure 6: {100} surfaces of aspirin, cleaved using the CERIUS® molecular modelling package.

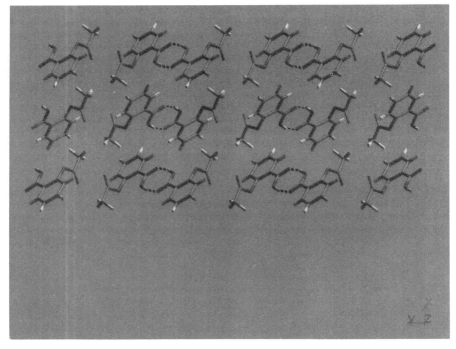

Figure 7: Cleaved {002} surfaces of aspirin.

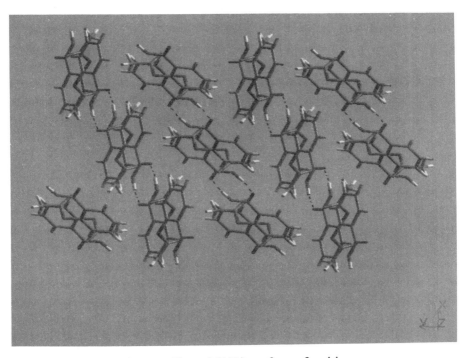

Figure 8: Cleaved {011} surfaces of aspirin.

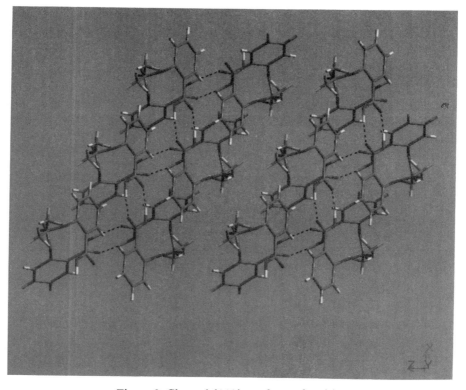

Figure 9: Cleaved {110} surfaces of aspirin.

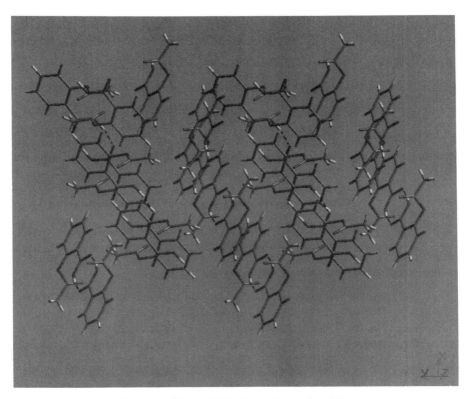

Figure 10: Cleaved {11-1} surfaces of aspirin.

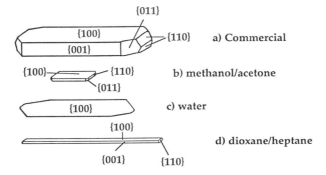

Figure 11: Variety of different growth habits of aspirin, obtained from different solvents. (Adapted from reference 19.) (a) "Commercial" aspirin, typically formed from a reaction crystallization from aromatic, acyclic or chlorinated solvents, (b) recrystallized from acetone or methanol solvents, (c) recrystallized from water, (d) recrystallized from dioxane or n-heptane.

morphology quoted *(19)* to be very similar to the domestic commercial product. Typically, aspirin is produced commercially from aromatic, acyclic or chlorinated hydrocarbons *(20)*, i.e. relatively non polar solvents in general. It would appear from the data given, that aspirin can exhibit solvent induced habit changes from both polar and non polar solvents.

The subject of current studies however, is the reconciliation between the predicted and growth habits via crystal chemistry considerations. It was reasoned that the differences in predicted morphological importance based on crystal chemistry considerations can explain the differences in morphological importance. However the {110}, {011}, {11-1} and {10-2} forms cleaved and viewed were postulated to have a higher charge density (thus leading to a higher growth rate), when theoretically compared to the {100} and {001} forms. Hence, one would then expect to see an increased morphological importance in the former faces in the presence of increasing solvent polarity, due to a reduction in growth caused by the crystal surface - solvent interactions. Experimentally, this does not appear to be so and these discrepancies are currently being investigated.

A method that may be applied to understand these discrepancies and why different morphologies are obtained from different solvents is the calculation of binding energies of the different solvents to the respective crystal faces and the ultimate effect on crystal morphology. This approach has been used previously *(21)* to good effect. This approach offers a route to investigate explicitly the role of crystal surface-solvent interactions and the impact on crystal morphology at the molecular level. This theoretical route will also be examined and an attempt to correlate the aforementioned crystallization studies with this method is currently underway.

Morphology/Materials Processing and Handling. There are a number of different relationships between morphology and solids processing and handling currently known in particle technology. For example, the relationship between tensile strength and morphology can easily be illustrated: dendritic crystals have a higher propensity to break during particle handling, leading to fines formation and a poorly defined particle size distribution. In the pharmaceutic industry, crystal morphology is known to have a major impact in tablet weight variation, tensile strength, dissolution rate and syringibility. Crystal habit can play a role in determining the filterability and flowability of a product *(22)*.

Looking at the predicted and growth morphologies, it is clear that the predicted morphologies and the commercially grown habit (typically crystallized from aromatic, acyclic or chlorinated hydrocarbons *(20)*) illustrated appear to be more prismatic and equidimensional than those recrystallized from different solvents (water, methanol and e.g., dioxane/heptane mixes) and hence offer greater ease of filtration, washing and are less subject to breakage. The other growth habits observed are not optimized for handling and it is clear from tensile strength considerations that the possibility of breakage with these habits is greater than those predicted or indeed, the commercial samples studied. However the samples recrystallized from different solvents form leaflike crystals *(19)* which have a lower tensile strength and are more likely to be subject to breakage. The crystals grown from dioxane/heptane and water are clear examples of such habits that are easily subject to breakage.

Conclusions

This paper has presented an initial study of the crystal morphology and crystal chemistry of aspirin. A cursory review of the effect of morphology on performance has also been presented. The predicted morphologies are in good agreement with commercially obtained samples - the {100} forms dominate, with the {011}, {001} and {110} forms present to a lesser extent. The predicted morphologies based on lattice energy calculations are somewhat elongated along the b direction, when compared to the geometrical BFDH model. The morphological importance of the simulations can be explained by hydrogen bonding, surface charge and steric considerations. There is a habit-solvent dependence noted in organic and inorganic solvents. The changes in morphology as yet cannot be reconciled with the crystal chemistry. It is hoped that current modeling studies coupled with experimentation will elucidate these effects.

Literature Cited

(1) Hartman, P. and Perdok, W.G. *Acta Cryst.* **1955** *8*, 49
(2) Berkovitch-Yellin, Z. *J. Amer. Chem. Soc.* **1985** *107*, 8239
(3) Docherty, R., Clydesdale, G., Roberts, K.J. and Bennema, P. *J. Phys. D. Applied Physics* **1991** *24*, 89
(4) Hartman, P. and Bennema, P. *J. Cryst. Growth* **1980** *49*, 145
(5) Bravais, A. *Etudes Crystallographiques, Paris* **1913**
(6) Friedel, G. *Bull. Soc. Fr. Mineral.*, **1907** *30* , 326
(7) Donnay, J.D.H. & Harker, D. *Am. Mineral.*, **1937** *22*, 463
(8) Jones, J.E. *Proc. Roy. Soc. Sect. A.* **1923** *106*, 441
(9) Lifson, S., Hagler, A.T. & Dauber P. *J.A.C.S.*, **1979** *101* , 5111
(10) Momany, F.A., Carruther, L.M., McGuire, R.F. & Scheraga, H.A. *J. Phys. Chem.*, **1974** *78* , 1579
(11) Nemethy, G., Pottle, M.S. & Scheraga H.A. *J. Phys. Chem.*, **1983** *87*, 1883
(12) Stewart, J.J.P. MOPAC - *A semiempirical Molecular Orbital Program; Journal of Computer -Aided Molecular Design*, **1990** *4*, 1
(13) Guest, M.F. & Kendrick, J. *GAMESS General Atomic And Molecular Electronic Structure* - **1986** University of Manchester Computing Center
(14) Kim, Y., Machida K., Taga, T. & Osaki, K. *Chem. Pharm. Bull.* **1985** *33*, 2641
(15) Clydesdale, G., Docherty, R. and Roberts, K.J. *Comp. Phys. Comm.* **1991** *64*, 311
(16) Molecular Simulations Inc., **1994** *CERIUS2*: version 1.6 , St. John's Innovation Center, Cambridge, U.K.
(17) Taylor, R. and Kennard, O. *Acc Chem. Res.* **1984** *17*, 320.
(18) Desiraju, G.R.. *Crystal Engineering - The Design Of Organic Solids*, Elsevier, Amsterdam. **1989**
(19) Watanabe, A., Yamaoka, Y., Takada, K. *Chem. Pharm. Bull.* **1982** *30*, 2958
(20) *Ullmann's Encyclopedia of Industrial Chemistry*, Ed. B. Elvers, VCH pub. Germany, **1993** *A23*, 481.
(21) Jang, S.M. and Myerson, A.S. *J. Crystal Growth* **1995** *156*, 459
(22) Palmer, R.C., and Batchelor, M.A. *Inst. Chem. End. Ind. Cryst. Symp.* **1969** *4*, 179

Chapter 2

Mechanism of Crystal Growth of Protein: Differential Scanning Calorimetry of Thermolysin Crystal Suspension

H. Ooshima, S. Urabe, K. Igarashi, M. Azuma, and J. Kato

Department of Bioapplied Chemistry, Osaka City University, 3–3–138 Sugimoto, Sumiyoshi-ku, Osaka 558, Japan

Mechanism of crystal growth of thermolysin was investigated through a thermal measurement by differential scanning calorimetry (DSC) of a crystal suspension which is under the progress of crystallization. The DSC curve obtained for the crystal suspension indicates that there are at least three groups of thermolysin differing in the thermal stability. The presence of the three groups is consistent with the two step mechanism of crystal growth of thermolysin previously proposed: the crystallization of thermolysin proceeds through two steps; the first step is the formation of primary particles with 20 nm in diameter; the second step is the crystal growth by highly ordered aggregation of the primary particles. In the DSC curve, Peak 1 (the denaturation temperature: 74.5-75.0 °C), Peak 2 (85.5-87.0 °C) and Peak 3 (95.5 °C) are assigned to the denaturation of monomers, primary particles (oligomers), and crystals, respectively.

Mechanism of protein crystallization has been investigated in several studies through various observations and analyses, for example by electron microscopy (*1, 2*), atomic force microscopy (*3, 4*), light interference (*5, 6*), light scattering (*7, 8*), small angle neutron scattering (*9, 10*), etc. Most of the investigations were made with the assumption that crystallization proceeds through nucleation and growth which takes place by attachment of monomeric protein molecules. However, some of them claim a participation of cluster and small particles in the crystal growth process. Niimura *et al* (*10*) proposed a novel mechanism for the crystal growth of lysozyme through a small angle neutron scattering study. When supersaturation is realized, two groups of aggregates have been created, i.e., Type I with a radius between 20 and 60 nm or more, and Type II with a radius of less than 4 nm; when nuclei for tetragonal crystal is formed, Type II starts to release monomers to the growing crystals. Type II particles themselves are not growth units of crystal. Growth unit is monomers released from Type II particles. The roll of Type I in the crystal growth is unknown. Kouyama *et al* (*4*) found, using atomic force microscopy on the crystal surface of bacteriorhodopsin, that spherical protein clusters with diameter of 50 nm are hexagonally close-packed. Electron microscopic observation of the spherical cluster revealed that the inside of the protein cluster is filled with mother liquor. The crystal is made up of hollow protein clusters. We (*1*) investigated the mechanism of crystallization of enzyme protein thermolysin and proposed the two step mechanism of crystal growth as illustrated in Fig.1. Crystallization of thermolysin proceeds through two steps; formation of primary particles (oligomers) with 20 nm in diameter and crystal growth by highly ordered aggregation of the primary particles. Ataka and Tanaka have reported the presence of an optimal initial supersaturation ratio to give a maximum in the growth rate and to give large crystals (*11*). We have reported a similar result for the crystallization of thermolysin (*12*). The result was well explained by the successive two step mechanism of crystallization.

According to the two step mechanism, in the crystallization solution there must exist at least three groups of molecules differing in the degree of intermolecular interaction, namely monomeric molecules, primary particles (oligomers), and crystals. In this work, differential scanning calorimetry (DSC) was carried out for the crystallization solution of thermolysin to confirm the two step mechanism.

EXPERIMENTAL

Materials.   Thermolysin from *Bacillus thermoproteolytic* was supplied from Daiwa Kasei Kogyo Co., Ltd. Since the enzyme powder supplied contains salts (12% sodium acetate and 24% calcium acetate) as stabilizer, it was purified by re-precipitation at pH 7.0 and the precipitate was freeze-dried at -40°C before use. The other chemicals used were of reagent grade.

Crystallization of thermolysin and differential scanning calorimetry (DSC) of the crystal suspension.   Supersaturated solution of thermolysin was prepared as follows. A given amount of the purified thermolysin was suspended in distilled water or aqueous solution of 20 vv% glycerol, and the pH was adjusted to 11.2 with 0.2N sodium hydroxide to completely dissolve it. After the solution was filtered with a 0.45 μm membrane filter to remove solid impurities, the pH was quickly adjusted to 7.0 with 0.2N acetic acid to induce crystallization. No buffer solution was used in the crystallization to avoid the effects of salts. The supersaturated solution of thermolysin was placed in a cell of 1.3 mL in a differential scanning calorimeter, Model MC-2, from MicroCal Co., USA. Crystals of thermolysin grew without agitation in the cell. The time at which the pH of the solution was adjusted to 7.0 was set to be zero in the crystallization time. The initial concentration of thermolysin was changed in the range of 3 to 15 mg/mL. All operations were carried out at 5 °C, and care was taken to prevent any contamination by dust.

After given crystallization time, the crystal suspension in the DSC cell (it should also contain molecules and oligomers) was heated from 5 °C to 105 °C at the rate of 10 or 90 °C/h.

Measurement of concentration of thermolysin during the crystallization and observation of denatured thermolysin precipitates by optical microscopy. Crystallization was also carried out in 2.0 mL glass tube to measure the concentration of thermolysin during the crystallization. Crystallization conditions were the same as those in the crystallization in the DSC cell.

The crystal suspension was filtered through a membrane filter of 0.45 μm (filtrate A). Part of the filtrate A was further filtered through an ultra filter (molecular mass cut off: 100 kD) (filtrate B). The filtrate A included particles of thermolysin smaller than 0.45 μm. The filtrate B mainly consisted of thermolysin molecules because the molecular mass of thermolysin is 34.6 kD. Both filtrates were analyzed by gel filtration high performance liquid chromatography (HPLC). In the analysis, the filtrates were adjusted to pH 11.0 before use as sample solutions to dissolve the crystals which had precipitated in the filtrates after the filtration.

The suspension of thermolysin crystal separately prepared in the glass tube was incubated at 85, 90, and 95 °C for 1 h. The thermally treated suspension was observed by optical microscopy.

RESULTS AND DISCUSSION

Figure 2 presents DSC curves for suspensions of thermolysin crystal (initial concentration: 3 mg/mL in water (initial supersaturation ratio: 18.7). The two suspensions are different in crystallization time, i.e. zero and 24 h. The scanning rate was 90°C/h. Although the curves were not smooth, maybe due to small signal/noise ratio which is caused by low concentration of thermolysin, at least two endothermic peaks which correspond to enthalpy changes by denaturation of thermolysin could be recognized. The curves suggest that there are at least two groups of thermolysin differing in thermal stability. Through the 24 h crystallization the peak at 87°C became small and the peak at 101°C became large. Figure 3 presents changes in the concentration of thermolysin during crystallization under the same conditions. Fractions I, II, and III correspond to thermolysin molecules, oligomers between 100 kD and 0.45 μm in diameter, and crystals larger than 0.45 μm, respectively. The only fraction which increases during crystallization is Fraction III. Therefore, the peak at 101°C shown in Fig.2 should be assigned to the denaturation of crystals. The peak at 87°C may be assigned the denaturation of monomers and/or the oligomers including small particles.

Since the DSC curves shown in Fig.2 are not very smooth, it is difficult to discuss details with respect to the assignment of the peaks. Figure 4 presents DSC curves for 15 mg/mL thermolysin after 24h crystallization. Temperature scan rates were 10 and 90°C/h. Only one peak was observed in each curve. The difference in the peak temperature between the two curves should be attributed to the fact that the heating rate of 90 °C/h is too fast to make the thermal denaturation at each temperature complete. In other words, it is caused by delay of denaturation which occurs at high scanning rate. Figure 5 presents change in the concentrations of monomers and oligomers less than 0.45 μm during crystallization of 15 mg/mL thermolysin. It is found that both of thermolysin molecules and oligomers are consumed almost completely by crystal growth at the initial stage of crystallization. It is due to the rapid crystal growth under high initial concentration of thermolysin. Therefore it can be concluded that both the peak observed at 101°C in the heating rate of 90 °C/h and the peak observed at 95.5 °C in the heating rate of 10 °C/h should be assigned to the denaturation of thermolysin crystal. The peak observed at 87°C, which was shown in Fig.2, was not observed here because of rapid crystal growth under high concentration of thermolysin.

In order to gain a sufficient concentration of Fraction I and II for DSC analysis, the crystallization was carried out under new conditions: high concentration of 15 mg/mL and high solubility, i.e., the crystallization was carried out in 20 vv% aqueous glycerol solution. Figure 6 presents changes in the concentration of thermolysin during the crystallization. High concentrations of Fractions I and II were obtained by crystallization within at least 8 hours. The slow scanning of DSC is not suitable for this sample, because Fractions I and II would decrease significantly during the temperature scanning. We employed only the higher scanning rate of 90 °C/h. Figure 7 presents changes in DSC curve obtained by changing the crystallization time from 0.5 to 8 h. When the crystallization time was 0.5 h, two peaks were observed. One was very small at 101 °C and the other peak observed at 93 °C was large. As the crystallization time becomes longer, namely as the crystal growth proceeds, the peak at 93 °C is getting smaller, and the peak at 101 °C is getting larger. This tendency is the same as that observed in Fig.2. The curve was deconvoluted to two peaks by assuming their denaturation temperature (top of the peak) to be 93 and 101 °C. The peak area was correlated to the concentration of Fractions I, II, and III. Figure 8 shows the results. The area of the peak at 93 °C

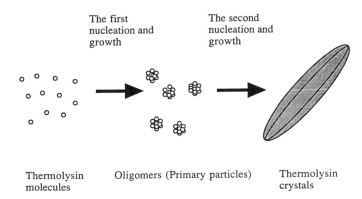

Fig.1 Two step mechanism of crystal growth of thermolysin (*1*).

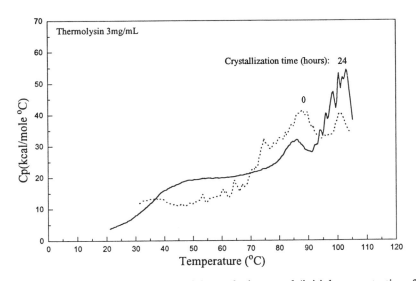

Fig.2 DSC curves for suspensions of thermolysin crystal (initial concentration: 3 mg/mL in water (initial supersaturation ratio: 18.7)) differing in crystallization time, i.e. zero and 24 h. Solvent: water (pH 7.0, 5 °C). Temperature scanning rate: 90°C/h.

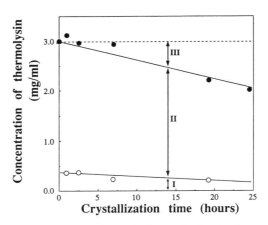

Fig.3   Changes in concentration of thermolysin during crystallization.
Crystallization conditions are the same as that shown in Fig.2. (●) filtrate A,
(O) filtrate B. Fractions I, II, and III: see the text.

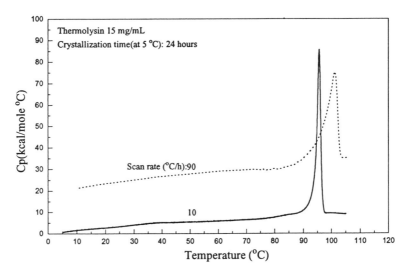

Fig.4 DSC curves for suspensions of thermolysin crystal (initial concentration: 15
mg/mL in water (initial supersaturation ratio: 93.7)). Crystallization time: 24 h.
Solvent: water (pH 7.0, 5 °C). Temperature scanning rate: 10 and 90°C/h.

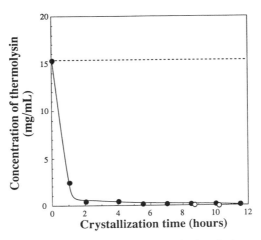

Fig.5 Changes in concentration of thermolysin during crystallization. Crystallization conditions are the same as that shown in Fig.4. (●) filtrate A, (O) filtrate B.

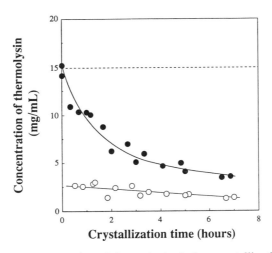

Fig.6 Changes in concentration of thermolysin during crystallization carried out in 20 vv% glycerol aqueous solution. Crystallization conditions: 15 mg/mL thermolysin, pH 7.0 and 5°C. (●) filtrate A, (O) filtrate B.

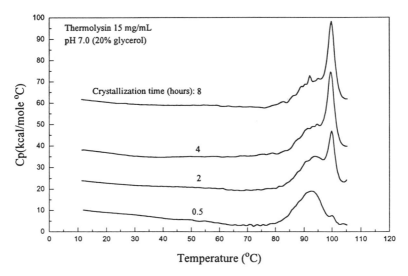

Fig.7 Changes in DSC curves during crystallization of 15 mg/mL-20 vv% glycerol thermolysin. Temperature scanning rate: 90°C/h.

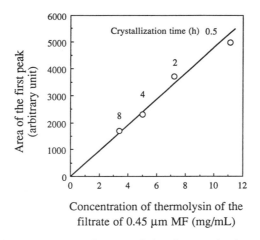

Fig.8 Correlation between peak area of the first peak shown in Fig.7 and summation of the concentration of monomers and oligomers of thermolysin.

increased in proportion to the summation of the concentration of Fraction I and II, neither of the each concentration of the two fractions. It means that the peak observed at 93 °C is assigned to the total of monomers and oligomers. We understand that we can distinguish the crystals from both of monomer and oligomers (primary particles) by DSC analysis. However, a question still remains. It is whether the peak at 93 °C might be composed of two peaks assigned to each fraction of I and II. The scanning rate of 90 °C/h may be too fast to recognize separately two denaturations whose denaturation temperatures are very close.

Another attempt to zoom up the thermal behavior of monomeric thermolysin and the oligomers was carried out. Two supersaturate solutions of thermolysin different in the initial concentration, that is 7 and 12 mg/mL, were prepared by using 20 vv% glycerol aqueous solution as solvent. The solutions were kept at 5 °C for 0.5 h in the DSC cell (that is, the crystallization time was 0.5 h), and DSC operation started at 10 °C/h of scanning rate immediately . Figure 9 presents the DSC curves. In the curve obtained for the sample of 7.0 mg/mL thermolysin, we can recognize at least two peaks as deconvoluted to Peak 1 (75.0 °C) and Peak 2 (85.5 °C). Figure 10 presents changes in the concentration of thermolysin during crystallization of 7.0 mg/mL thermolysin at pH 7.0 and 5 °C. It indicates that formation of large crystals (Fraction III) is slow and a large portion of thermolysin is in the state of oligomers. The situation hardly changes even after incubation for 24 h. Namely, in the sample there are only monomers and oligomers. Furthermore the main component is oligomers. From this observation Peaks 1 and 2 are assigned to monomers and oligomers, respectively. On the other hand, in the case of 12 mg/mL thermolysin, we visibly observed many small crystals after the incubation of 1.5 h, although data on changes in the concentration were not taken. As we would expect from the presence of crystals, the DSC curve of 12.0 mg/mL thermolysin could be deconvoluted to three peaks including Peaks 1 and 2. Since the denaturation temperature of Peak 3 is the same as that observed in Fig.4, the peak is assigned to the denaturation of crystals.

The microscopic observation supports the above-mentioned assignment. Thermolysin crystals in 20 vv% glycerol aqueous solution was incubated at 85, 90, and 95 °C for 1 h. As Fig.11 shows, when the crystals were incubated at 85 °C, no visible change in the crystal shape and size occurred. At 90 °C, a small portion of crystals collapsed and changed to white precipitate, but most of crystals were keeping the original shape. At 95 °C, all crystals completely collapsed and changed to white precipitate. It was found from these observations that Peaks 1 (74.5-75.0 °C) and 2 (85.5-87.0 °C) are phenomena independent of crystals, i.e. the denaturation of monomeric thermolysin molecules and oligomers. We also confirmed that Peak 3 (95.5 °C) corresponds to the denaturation of crystals.

The presence of three groups of thermolysin differing in the thermal stability is consistent with the two step mechanism of the crystal growth previously proposed (*1*) (see Fig.1). The thermal stability of thermolysin increases through formation of the primary particle (oligomers) and furthermore through formation of crystals. The difference in the thermal stability must be caused by the difference in conformation of thermolysin molecules which in turn must be caused by the differences in intermolecular interactions. This consideration leads to the conclusion that although the crystals of thermolysin grow through the highly ordered aggregation of the primary particles, the feature of the intermolecular interactions in the primary particles is not strictly the same as that in the crystal.

CONCLUSION

The thermal analysis by DSC was carried out for thermolysin crystal suspensions in which crystal growth was on progress. As a result, it was found that three groups of thermolysin differing in the thermal stability exist. Each group was assigned to monomers, oligomers, and crystals, respectively. The denaturation temperature of

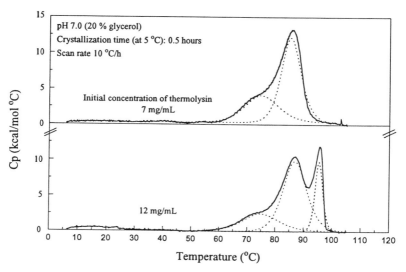

Fig.9 DSC curves for suspensions of thermolysin crystal. Initial concentration of thermolysin: 7 (above) and 12 mg/mL (below). Solvent: 20 vv% glycerol aqueous solution (pH 7.0, 5 °C). Crystallization time: 0.5 h. Temperature scanning rate: 10°C/h.

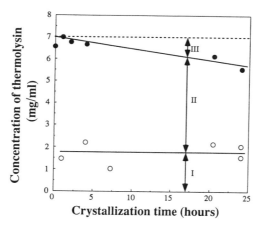

Fig.10 Changes in concentration of thermolysin during crystallization of 7 mg thermolysin/mL-20 vv% glycerol aqueous solution. (●) filtrate A, (O) filtrate B.

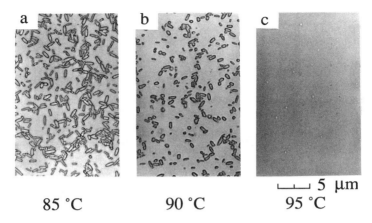

85 °C   90 °C   95 °C

Fig.11 Microscopic observation of thermolysin crystal thermally pretreated for 1 h at 85, 90 and 95 °C.

monomers, of the primary particles (oligomers) and of the crystals are 74.5-75.0, 85.5-87.0 and 95.5 °C, respectively. The presence of these three groups of thermolysin is consistent with the two step mechanism of the crystal growth previously proposed (*1*).

Acknowledgment

The authors would like to thank Daiwa Kasei Kogyo Co., Ltd., for supplying thermolysin.

References

1. Sazaki, G.; Ooshima, H.; Kato, J.; Harano, Y.; Hirokawa, N. *J. Crystal Growth* **1993**, *130*, 357.
2. Durbin, S. D.; Feher, G. *J. Mol. Bio.* **1990**, *212*, 763.
3. Durbin, S. D.; Carlson, W.E. *J. Crystal Growth* **1992**, *122*, 71.
4. Kouyama, T.; Yamamoto, M.; Kamiya, N.; Iwaki, H.; Ueki, T.; Sakurai, I. *J. Mol. Biol.* **1994**, *236*, 990.
5. Komatsu, H.; Miyashita, S.; Suzuki, Y. *Jpn. J. Appl. Phys.* **1993**, *32*, 1855.
6. Miyashita, S.; Komatsu, H.; Suzuki, Y.; Nakada, T. *J. Crystal Growth* **1994**, *141*, 419.
7. Azuma, T.; Tsukamoto, K.; Sunagawa, I. *J. Crystal Growth* **1989**, *98*, 371.
8. George, A.; Wilson, W.W. *Acta Cryst.* **1994**, *D50*, 361.
9. Niimura, N.; Minezaki, Y.; Ataka, M.; Katsura, T. *J. Crystal Growth* **1994**, *137*, 671.
10. Niimura, N.; Minezaki, Y.; Ataka, M.; Katsura, T. *J. Crystal Growth* **1995**, *154*, 136.
11. Ataka, M.; Tanaka, S. *Biopolymers* **1986**, *25*, 337.
12. Sazaki, G.; Aoki, S.; Ooshima, H.; Kato, J. *J. Crystal Growth* **1994**, *135*, 199 .

Chapter 3

Source of Nuclei in Contact Nucleation as Revealed by Crystallization of Isomorphous Alums

Manijeh M. Reyhani and Gordon M. Parkinson

A. J. Parker Cooperative Research Centre for Hydrometallurgy,
School of Applied Chemistry, Curtin University of Technology,
GPO Box U 1987, Perth 6001, Western Australia

Potash alum, $KAl(SO_4)_2.12H_2O$, and chrome alum, $KCr(SO_4)_2.12H_2O$, crystallize from their respective aqueous solutions as regular octahedra in almost identical forms and are known to be isomorphous. Alum crystals are also easily formed by secondary crystallization. Gentle crystal contact between a parent alum crystal and a solid surface under supersaturated aqueous solution produces a large number of secondary nuclei. In this study, the formation of secondary nuclei of potash alum crystals in its supersaturated solution by contacting a known crystal surface of chrome alum with a TEM grid, and, conversely, the formation of chrome alum crystals in its supersaturated solution by contacting a known crystal surface of potash alum are reported. These experiments have been used to study closely the source of nuclei in contact nucleation. The results obtained using analytical transmission electron microscopy will be discussed.

Contact nucleation is probably the most important source of secondary nuclei in an industrial crystallizer (1-4). New crystals are formed due to the prior presence of other growing crystals. The source of nuclei in contact nucleation has been studied by several investigators and two major theories have been proposed (5-11). One assumes that the source of nuclei is an ordered, intermediate layer of solute adjacent to a growing crystal surface (5-8) and the other suggests the origin of the secondary nuclei is from the parent crystal, involving the generation of new particles by microattrition of a growing crystal surface (9-11).

Potash alum, $KAl(SO_4)_2.12H_2O$, and chrome alum $KCr(SO_4)_2.12H_2O$ crystallize from their respective aqueous solutions as regular octahedra in almost identical forms and are known to be isomorphous. Alum crystals are also easily formed by secondary nucleation. The objective of this work is to study more closely the source of nuclei in contact nucleation by making use of the isomorphous relationship between the crystal structures of potash and chrome alums, and the power of the analytical transmission electron microscope (TEM) to differentiate between very small volumes of the chemically different components.

Experimental

Two aqueous systems have been studied: pure (99.5%) aluminium potassium sulphate, $KAl(SO_4)_2.12\ H_2O$, (potash alum-water) system and pure (99.6%) chromium aluminium sulphate, $KCr(SO_4)_2.12\ H_2O$, (chrome alum-water). A 3 mm diameter, 200 mesh copper grid coated with carbon film was brought into gentle contact with specific growth faces of a potash alum crystal in a supersaturated solution of chrome alum and conversely with a chrome alum crystal in a supersaturated solution of potash alum. Immediately after contact, the grids were removed from the solution, and the excess liquid was rapidly drained off. Nuclei were produced in the range of 50nm-5μm and were studied using a Philips 430 analytical transmission electron microscope, and analysed by electron diffraction and energy dispersive X-ray analysis (EDX).

Results

Figure 1 shows the morphology of a typical alum crystal (12). Three distinctive crystal faces of (100), (110) and (111) are shown in this figure. Potash alum and chrome alum crystals are isomorphous and both exhibit this structure. Both materials form smooth overgrowths onto crystals of each other in supersaturated solutions. Production of secondary nuclei by contact nucleation in potash alum has been previously reported in a number of papers, as it is an ideal system for such studies (9, 13). Contacting an identified surface of a potash alum crystal, for example the (100), (110) and (111) growth faces with a solid surface in a supersaturated solution of the same alum resulted in the production of many secondary nuclei of an identical orientation (13). Similar experiments have been carried out in this work by choosing a chrome alum solution when the potash alum parent crystal is used and a potash alum solution when a chrome alum parent crystal is used. Figure 2a shows one of the crystals formed by contacting a (111) face of a chrome alum crystal with a TEM grid in a supersaturated potash alum solution. As shown in figure 2b, an EDX spectrum taken from this crystal shows no trace of chromium, indicating that there has been no transfer of solid in the formation of the new crystal. The equivalent experiment, ie, contacting potash alum crystal in a supersaturated solution of chrome alum, resulted in the production of chrome alum crystals without any trace of aluminium coming from the solid crystal.

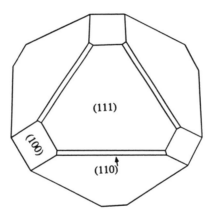

Figure 1　Schematic diagram of the observed morphology of alum crystals showing different crystal faces.

Figure 2a   Transmission electron micrograph, showing a bright field image of a secondary nucleus produced by contact with a (111) face of a chrome alum parent crystal.

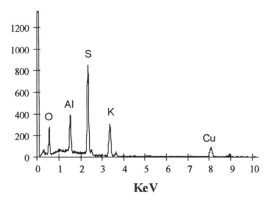

Figure 2b   An EDX spectrum taken from the secondary nucleus produced by contact with a (111) face of a chrome alum parent crystal.

Examination of the very fine particles produced by this method (for example, those of 150 nm in size, shown in figures 3a & 3b) also showed no trace of solid transfer from the contacted crystal. This is important because at such small particle sizes the presence of any embedded material would not be obscured by the absorption of emitted X-rays. These nuclei were produced in a potash alum solution by contacting a chrome alum (111) crystal face (figure 3a) and in a chrome alum solution by contacting a potash alum (100) face (figure 3b). Figures 4a & 4b show the EDX spectra taken from these particles. Although the small nuclei do not show the external form of the crystal orientation as clearly as is observed at lower magnification with larger crystals (figure 2a), nonetheless they do exhibit the same orientation, as revealed by electron diffraction, which also indicates that they have a well ordered crystalline structure.

Discussion

The above results favour the proposition that the intermediate ordered layer of solute adjacent to a growing crystal surface is the source of secondary nuclei in this case (5-8). However, there is the possibility that some transfer of solid material may have occurred and it has not been detected. This could be due either to the small size of the fragment transferred or to the quick overgrowth of the parent crystal in supersaturated solution; ie, if a sufficiently thick overgrowth of the counter material from solution occurred prior to contact, then any solid transferred would not contain the characteristic element from the source crystal, and hence would not be detected. The degree of supersaturation and control of the rate of growth over the parent crystal are thus very important factors. To clarify this, potash alum crystals, which have lower solubility than chrome alum, have been used as parent crystals in a range of chrome alum solutions that were undersaturated and supersaturated. Results show that in the case of an undersaturated solution of chrome alum, no nuclei of potash alum are formed from the solid crystal, whereas nuclei of chrome alum were detected from both saturated and supersaturated solutions of chrome alum by contacting the parent crystal of potash alum crystal. Comparing these results indicates that the secondary nuclei that are produced truly come from the surface between the original crystal and the solution and not from a thick overgrowth (14).

Acknowledgments

This work has been supported under the Australian Government's Cooperative Research Centres Programme, and this support is gratefully acknowledged.

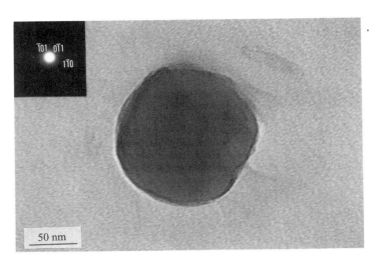

Figure 3a   Transmission electron micrograph showing a bright field image of a small secondary nucleus produced by contact with a (111) face of a chrome alum parent crystal in a supersaturated potash alum solution.

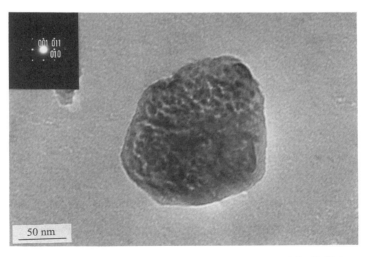

Figure 3b   Transmission electron micrograph showing a bright field image of a small secondary nucleus produced by contact with a (100) face of a potash alum parent crystal in a supersaturated chrome alum solution.

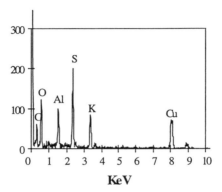

Figure 4a An EDX spectrum taken from the small secondary nuclei produced by contact with a (111) face of a chrome alum parent crystal in a supersaturated potash alum solution (see figure 3a).

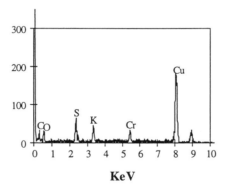

Figure 4b An EDX spectrum taken from the small secondary nuclei produced by contact with a (100) face of a potash alum parent crystal in a supersaturated chrome alum solution (see figure 3b).

Literature Cited

1. E. G. Denk, Jr. and G. D. Botsaris, J. of Crystal Growth **1972,** 13/14, 493.
2. J. Mathis-Lilley and K. A. Berglund, AIChE J. **1985,** Vol. 31, No.5, 865.
3. M. Liang, R. W. Hartel and K. A. Berglund, J. Engineering Science **1987** Vol. 42, No.11, 273.
4. M. K. Cerreta and K. A. Berglund, J. of Crystal Growth **1990,** 102, 869.
5. E. C. Powers, Nature **1956,** 178, 139.
6. K. A. Berglund, M. A. Larson, AIChE Symposium Series, **1982** 9.
7. P. Elankovan and K. A. Berglund, Applied Spectroscopy**1986**, Vol. 40, No. 5, 712.
8. P. Elankovan and K. A. Berglund, AIChE J, **1987,** Vol.33, No.11, 1844.
9. J. Garside, and M. A. Larson, J. of Crystal Growth **1978,** 43, 694.
10. R. Wissing, M. Elwenspoek and B. Degens, J. of Crystal Growth **1986,** 79, 614.
11. K. Shimizu, K. Tsukamoto, J. Horita and T. Tadaki, J. of Crystal Growth **1984,** 69, 623.
12. N. Sherwood and T. Shripathi, Faraday Discussion **1993.**, 95, 173.
13. M. M. Reyhani and G. M. Parkinson, J. of Crystal Growth, in press, **1996**.
14. M. M. Reyhani and G. M. Parkinson, to be published.

Chapter 4

A Kinetic Model for Unsteady-State Crystal Growth in the Presence of Impurity

N. Kubota, M. Yokota, and L. A. Guzman[1]

Department of Applied Chemistry and Molecular Science, Iwate University,
4–3–5 Ueda, Morioka 020, Japan

A kinetic model explaining the unsteady state impurity action for the crystal growth is presented. The unsteady state behavior, which is related to the gradual decrease of crystal growth rate with time in the presence of impurity, is attributed to the non-equilibrium adsorption kinetics of the impurity. The slow adsorption process is described by the non-equilibrium Langmuir adsorption mechanism. Equations for theoretical growth rate and growth length of crystal are developed. Literature growth data for various systems are qualitatively analyzed by the model. The model is reduced to the equilibrium adsorption model [N. Kubota and J. W. Mullin, *Journal of Crystal Growth*, *152*(1995)203], if the impurity adsorption proceeds rapidly to reach equilibrium state. The effects of supersaturation and temperature on the impurity action are assessed in the light of an impurity effectiveness factor, α.

Small amount of impurities sometimes retard dramatically the crystal growth rate. Chromium(III) , for example, suppresses the crystal growth of potassium sulfate(*1*), ammonium dihydrogen phosphate(*2*) and ammonium sulfate (*3*), etc. in aqueous solutions. Other metallic ions, iron(III) , aluminium(III) are also effective impurities (*4*).

In some cases, the crystal growth rate in the presence of impurity decreases gradually over several tens of minutes after an impurity is introduced into the system, and finally reaches a steady state value or zero value. Experimental data of this unsteady state impurity action have been reported fragmentarily in the literature (*2, 3, 5*). But no theoretical explanation has been given so far.

In this paper, a kinetic model (non-equilibrium adsorption model) will be presented to explain the unsteady state impurity action. Literature data of the unsteady state impurity action (*1-3 ,5*) will be also shown and compared with the model.

Non-Equilibrium Adsorption Model

The relative face growth rate, G/G_0, is assumed to be written by the following linear function of the surface coverage, θ, by impurity species,

[1]Current address: Department of Chemical Engineering, Himeji Institute of Technology, 2167 Shosha, Himeji, Hyogo 67–22, Japan

$$\frac{G}{G_0} = 1 - \alpha\theta \tag{1}$$

where α is the impurity effectiveness factor (6) and G_0 the growth rate for the pure system. The factor, α, is a parameter accounting for the effectiveness of an impurity under a given growth condition (temperature and supersaturation). It does not change with time, while the surface coverage is assumed to change.

If the Langmuir (non-equilibrium) adsorption mechanism applies, the net adsorption rate, i. e., the increasing rate of the surface coverage, is given as the difference between the adsorption and desorption rates by the following equation,

$$\frac{d\theta}{dt} = k_1(1-\theta)c - k_2\theta \tag{2}$$

where t is time , k_1 and k_2 are adsorption and desorption rate constants, respectively. Under the condition of constant impurity concentration, c, integrating Equation 2 with the initial condition of $\theta = 0$ at $t = 0$ gives the following equation,

$$\theta = \theta_{eq}(1-exp(-\frac{t}{\tau}) \tag{3}$$

The surface coverage increases exponentially with time and reaches a final equilibrium value θ_{eq}, which is given by the Langmuir adsorption isotherm,

$$\theta_{eq} = \frac{Kc}{1 + Kc} \tag{4}$$

where K is the Langmuir constant $(=k_1/k_2)$. The rate of adsorption is governed by a time constant, τ, which is given by

$$\tau = \frac{1}{(k_1c + k_2)} \tag{5}$$

Time changes of the relative surface coverage, in theoretical adsorption processes, are shown in Figure 1 for two cases of $\tau = 1$ min and 30 min. The adsorption is seen to proceed very slowly for the case of $\tau = 30$ min.

Inserting Equation 3 into Equation 1, the relative growth rate, G/G_0, is given as

$$\frac{G}{G_0} = 1 - \alpha\theta_{eq}[1- exp(- \frac{t}{\tau})] \tag{6}$$

If the time constant τ is large, the relative growth rate decreases slowly. The slow adsorption process is the cause of the slow growth rate change (slow unsteady state impurity action). Figure 2 shows theoretical growth rates as a function of dimensionless time, t/τ , for different values of $\alpha\theta_{eq}$. For weak impurities ($\alpha < 1$), the relative growth never reaches zero since θ_{eq} is not larger than unity. It approaches zero asymptotically at $t = \infty$ when $\alpha = 1$. For strong impurities ($\alpha > 1$), the relative growth rates is able to reach zero at the characteristic time, t_c, if $\alpha\theta_{eq} > 1$. The characteristic time, t_c, is given by θ

$$t_c = \ln(\frac{\alpha\theta_{eq})}{\alpha\theta_{eq} - 1})\tau \tag{7}$$

Integration of Equation 6 gives the growth length of crystal, ΔL,

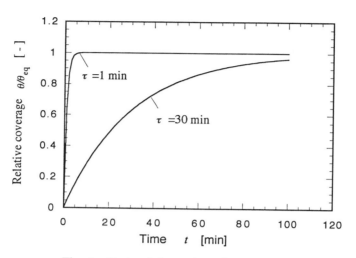

Fig. 1   Fast and slow adsorption processes

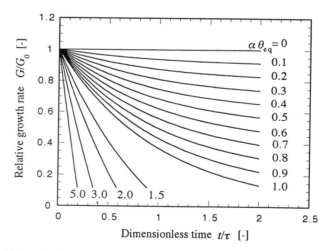

Fig. 2   Theoretical relative growth rates as a function of dimensionless time
for different values of $\alpha\theta_{eq}$

$$\frac{\Delta L}{G_0 \tau} = (1 - \alpha\theta_{eq})(\frac{t}{\tau}) + \alpha\theta_{eq}[1 - exp(-\frac{t}{\tau})] \tag{8}$$

In this equation the growth length, ΔL, is given in a dimensionless form. Figure 3 shows Equation 8 with different $\alpha\theta_{eq}$ values. Equation 8 is always valid if $\alpha\theta_{eq}$ is not larger than unity. While, if $\alpha\theta_{eq} > 1$, it is valid only up to the characteristic time t_c. After that, the dimensionless growth length becomes constant and its value is given as

$$\frac{\Delta L}{G_0 \tau} = 1 - \frac{exp(-\frac{t_c}{\tau})}{1 - exp(-\frac{t_c}{\tau})}(\frac{t_c}{\tau}) \tag{9}$$

Equation 9 gives the locus line of the characteristic time, t_c (see Figure 3).

A qualitative Comparison with the literature data

The reported data on impurity action are scattered. There is no generalized unified approach to assess the effect of impurity on growth. Some of them are introduced and qualitatively analyzed with the aid of the model described above.

Growth Rate of Potassium Sulfate Crystal in the Presence of Chromium(III) (1). The growth length of the {110} faces of potassium sulfate crystal was measured in the presence of chromium(III) in a flow cell, where the flow rate of the solution was adjusted to a level in the surface-integration-controlled regime. Chromium(III) was added as $Cr_2(SO_4)_3 \cdot xH_2O$. [The hydration number x was 9.] The results are shown in Figure 4, where the normalized growth length is plotted with growth time for different (impurity) chromium(III) concentrations. The normalized growth length is defined as the ratio of actual growth length in a contaminated solution, ΔL, with the total growth length in the chromium(III)-free solution during the total growth time, $G_0 t_G$. It reaches unity for a chromium(III)-free system at the end of run.

The trivalent chromium(III) is a very strong impurity. Five ppm of chromium(III) is sufficient to completely stop the growth process of potassium sulfate crystal. Typical unsteady state characteristic of chromium(III) can be clearly seen at 2 ppm concentration. On the whole, the empirical growth length behavior is similar to that of model predictions as demonstrated in Figure 3. [Note: The parameter, $\alpha\theta_{eq}$, in Figure 3 corresponds to the impurity concentration, since θ_{eq} is given as a function of the impurity concentration by Equation 4 while α is constant for the given system at a given growth condition as given by Equation 11 later.)

Growth Rate of Ammonium Sulfate Crystal in the Presence of Chromium(III) (3). The growth rate of ammonium sulfate crystals was measured in the presence of chromium(III) in a flow cell. Chromium(III) was added as $Cr_2(SO_4)_3$ [the number of water of the hydrate was not indicated in the original paper]. The results are shown in Figure 5, where the growth length, ΔL, was directly plotted as a function of growth time for different impurity concentrations.

As manifested in the growth of potassium sulfate (Figure 4), here also chromium(III) is very effective in suppressing the growth rate of ammonium sulfate crystal. The unsteady state impurity action is evident at various chromium(III) concentrations.

Growth Rate of Urea in the Presence of Biuret (5). Figure 6 reports the unsteady state growth behavior of urea crystal in the [001] direction in the presence of biuret

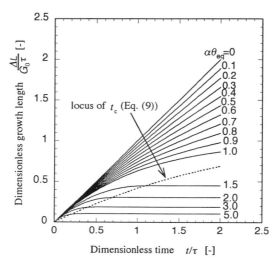

Fig. 3   Dimensionless growth length as a function of dimensionless time for different values of $\alpha\theta_{eq}$

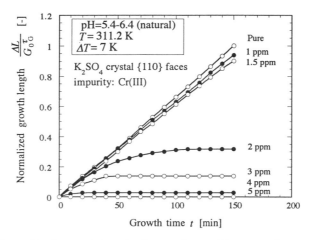

Fig. 4   Normalized growth length of potassium sulfate crystal as a function of time

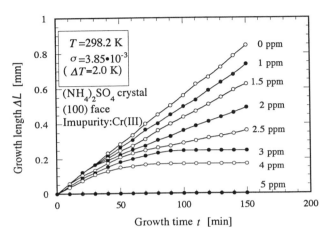

Fig. 5 Growth length of ammonium sulfate crystal as a function of time

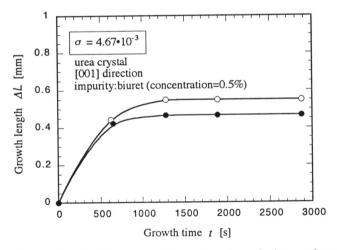

Fig. 6 Growth length of urea crystal as a function of time - plots for two crystals growing under the same condition

as an impurity. The growth rate (slope of the lines) decreases with time and reaches zero in 1200 s (20 min). This is also an example of unsteady impurity action.

Growth Rate of Ammonium Dihydrogen Phosphate (ADP) Crystal in the Presence of Chromium(III) (2). An ADP crystal exhibits the unsteady state growth behavior in the presence of chromium(III) added as $CrCl_3 \cdot 6H_2O$ (Figure 7). The relative growth rate (not the growth length as in the previous figures) decrease gradually with time. The theoretical equation for the relative growth rate (Equation 6) was fitted to the data by using a non-linear least squares method (solid line in the figure). There is close match between data points and theoretical curve. The parameters obtained by fitting are shown in the figure.

Condition for the Unsteady State Impurity Action.   Four cases of the unsteady state growth behavior have been shown. Every case was for strong additive ($\alpha > 1$), since the characteristic time t_c was existing for all the cases. Therefore, strong additive could be one of the necessary conditions. Meanwhile, it is theoretically concluded that the time constant for the adsorption of an impurity, τ, should be large for unsteady state behavior to occur (see Equation 3). According to Equation 5, a large time constant, τ, can be realized only when the impurity concentration, c, is low. This requirement from the theory corroborates the above speculation that large effectiveness factor, α, is a necessary condition, since θ_{eq} could be small for that case and, hence, the impurity concentration, c, could be low. Measurements of adsorption kinetics are necessary to clarify the condition for the unsteady state impurity action.

Physical Significance of Impurity Effectiveness Factor, α (6)

The impurity effectiveness factor, α, plays an important role in the assessment of impurity action. Its physical meaning is discussed here briefly.

The non-equilibrium adsorption model, mentioned above, is reduced to a model proposed previously by Kubota and Mullin (6) , where equilibrium adsorption of an impurity is assumed. [It is named equilibrium adsorption model.]   As τ approaches zero, Equation 6 reduces to,

$$\frac{G}{G_0} = 1 - \alpha\theta_{eq} \tag{10}$$

This is the equation for the equilibrium adsorption model. As discussed in the previous paper, if impurity species adsorb only on the step lines and the pining mechanism (7) applies for the impurity action, the following simple equation can be derived (6).

$$\alpha = \frac{f\gamma a}{kT\sigma L} \qquad (\sigma \ll 1) \tag{11}$$

where f is a factor to take into account for stereochemical effect of an impurity, γ the edge free energy, a the size of crystallizing species, k the Boltzmann constant, T the temperature, σ the supersaturation and L distance between the active sites for impurity to adsorb on the step lines.

In Figure 8, the effect of supersaturation on the effectiveness factor, α, for raffinose (impurity) in the growth of sucrose from aqueous solution is shown. Data points in the figure were calculated from the growth data reported by Albon and Dunning (8). The effectiveness factor, α, is seen to be inversely proportional to the supersaturation, as it is predicted by Equation 11.

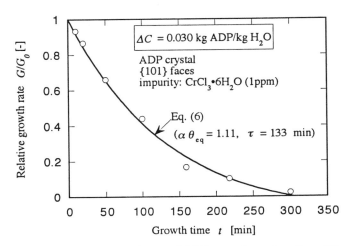

Fig. 7 The relative growth rate of ADP crystal as a function of time

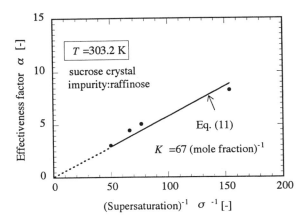

Fig. 8 Impurity effectiveness factor vs. supersaturation

Legend of Symbols

a : size of growth unit or area per growth unit appearing on the surface, m^2

C : concentration of crystallizing species, mol/dm^3 or kg/kg H$_2$O

C_s : saturated concentration of crystallizing species, mol/dm^3 or kg/kg H$_2$O

ΔC: suppersaturation $(=C-C_s)$, kg/kg H$_2$O

c : impurity concentration, ppm (mg/dm^3) or mole fraction

f : factor for stereochemical effect of impurity

G : growth rate in the presence of impurity, m/s or mm/min

G_0: growth rate of crystal in pure solution, m/s or mm/min

k : Boltzmann constant, J /K

k_1 : adsorption constant (Equation 2), (mg/dm^3)$^{-1}$·s^{-1} or (mole fraction)$^{-1}$·s^{-1}

k_2 : desorption constant (Equation 2), s^{-1}

K : Langmuir constant $(=k_1/k_2)$, (mg/dm^3)$^{-1}$ or (mole fraction)$^{-1}$

L : average distance between active sites on the step lines, m

ΔL: growth length of crystal, m or mm

T : temperature, K

t : growth time, s or min

t_c : characteristic time of unsteady state impurity action (Equation 7), s

t_G : total growth time, min

α : impurity effectiveness factor

θ : coverage of active sites of crystal surface by adsorbed impurities

θ_{eq}: θ at equilibrium

γ : edge free energy of step, J/m

σ : relative supersaturation $(=C/C_s)$

τ : time constant of adsorption, s or min

Literature Cited

(1) Guzman, A. L.:*Master thesis (Iwate University)*, March, **1995**

(2) Mullin, J. W.: Davey, R. :*Journal of Crystal Growth*, **1974**, *23*, pp. 89-94

(3) Kitamura, M.: Ikemoto, K.: Kawamura, Y.: Nakai, T. :*Kagakukogaku Ronbunshu*, **1990**, *16*, pp. 232-238

(4) Mullin, J. W. : *Crystallization*, 3rd. Butterworth-Heinemann: London, **1993**, p 255

(5) Davey, R. : Fila, W.: Garside, J.: *Journal of Crystal Growth*, **1986**, *79*, pp. 607-613

(6) Kubota, N.: Mullin, J. W.:*Journal of Crystal Growth* **1995**, *152*, pp. 203-208

(7) Cabrera, N.: Vermilyea, D. A.: in *Growth and Perfection of Crystals*; Doremus, R. H.: Roberts, B. W.: Turnbull, D. , Ed.; Wiley: New York , **1958**, pp 393-410

(8) Albon, N. : Dunning, W. J.: *Acta Crysta.*, **1962**, *15*, pp. 474-476

CRYSTALLIZATION OF OPTICAL ISOMERS AND POLYMORPHS

Chapter 5

Concentration Change of L-SCMC [S-(Carboxymethyl)-L-cysteine] in a Supersaturated Solution of DL-SCMC with Suspended L-SCMC Seeds in a Batch Operation

Ken Toyokura, M. Kurotani, and J. Fujinawa

Department of Applied Chemistry, Waseda University, 3–4–1 Okubo, Shinjuku-ku, Tokyo 169, Japan

Changes of the concentration of L-SCMC and D-SCMC in a supersaturated solution of DL-SCMC in which L-SCMC seed crystals were suspended, were observed in batch tests. The rate of the concentration decrease of L-SCMC by the growth of L-SCMC seeds was correlated with supersaturation and the amount of L-SCMC seeds in two different solutions characterized by dominant component of D-SCMC and L-SCMC, respectively. Operational lines expressed by concentration of D-SCMC and L-SCMC were observed and bending points of the lines were correlated with the induction time defined by the time that the concentration of D-SCMC started to decrease by growth of D-SCMC nuclei born on the surface of L-SCMC seeds. The induction time observed was also correlated with supersaturation and the amount of L-SCMC seeds added. Operational conditions for a proposed optical resolution process by crystallization were discussed on an induction time so that L-SCMC seed in a supersaturated DL-SCMC solution with NaCl might grow without any inclusion of D-SCMC, and a process for deciding the volume of a crystallizer for optical resolution was proposed on the basis of the data obtained in this study.

Crystallization of L-SCMC (S-Carboxymethyl-L-cysteine) from DL-SCMC solution, which sodium chloride is contained, has been studied for the purpose to develop an optical resolution process by crystallization. In a crystallization process, crystal size and purity and product amount of crystal are important. Growth rates of L-SCMC seeds and surface nucleation of D-SCMC on the surface of L-SCMC seed crystals, which affect crystal size and purity, were reported by Yokota et al. (1). Using an optical resolution process, the product amount of L-SCMC crystal was studied by the decrease of the concentration of L-SCMC in a DL-SCMC solution. In this study, the concentration change of D-SCMC and L-SCMC in a DL-SCMC supersaturated solution, in which L-SCMC seeds were suspended, were observed. The decreasing rate of the concentration of L-SCMC was different depending on whether the solution was classified by dominant component of L-SCMC or D-SCMC. Particular correlative equations were obtained for either case. The operational conditions at which an L-SCMC seed crystal grew without crystallization of D-SCMC were studied by the induction time for surface nucleation of D-SCMC and a method for deciding the volume of crystallizer for optical resolution process was discussed based on the data obtained in this study.

Experiments.

A racemic DL-SCMC solution saturated at 308K was fed to a 500ml crystallizer and kept with agitation at 10K higher than the saturated temperature of the solution. A certain amount of L-SCMC crystals were added to the agitated solution and dissolved completely in it. This DL-SCMC solution, in which the initial supersaturation of L-SCMC ΔC_{L0} was higher than that of D-SCMC ΔC_{D0}, was cooled down at a constant rate of 0.6 K/min to the temperature T_{D0}. When the temperature of the solution reached T_{D0}, 2.5 to 34g of L-SCMC seed crystals whose size was 1100 to 2000μm, were fed into the solution. The seed crystals of L-SCMC used were made by growth of nuclei born in a stationary solution of supersaturated L-SCMC. 10ml of slurry was sampled and the solution was separated through a 0.45μm membrane filter. Concentration of L- and D-SCMC of the sampled solution were analyzed by HPLC (High Performance Liquid Chromatography), and concentration change of the solution was determined with time. Experiments were carried out batchwise and operational conditions are shown in Table I. ΔT_{D0} is the initial supercooling of the solution which was expressed by the difference between the operational temperature T_{D0} and the saturation temperature for D-SCMC. The revolution rate of an agitator in this study was 200r.p.m..

Results and Discussion.

Change of the concentrations of L- and D-SCMC in an operating solution.
Typical examples of observed concentration changes of L- and D-SCMC are shown in Figure 1, for Runs.1-c and 4-g, whose operational conditions are shown in Table I. The concentration of L-SCMC decreased rapidly just after L-SCMC seed crystals were added, and then gradually decreased because of the growth of L-SCMC seed. The concentration of D-SCMC in the solution was not changed for a while and then started to decrease. The induction time which was defined as the elapsed time till the concentration of D-SCMC start to decrease, was affected by the initial supercooling of D-SCMC component. The reason of decrease of D-SCMC concentration in Figure 1 can be easily understood with Yokota's report which explain that surface nucleation of D-SCMC on L-SCMC seed crystals in DL-SCMC supersaturated solution took place after some induction time which is affected by supersaturation (2). The effect of the amount of seeds on the decreasing rate of L- and D-SCMC concentration is shown in Figure 2. In this study, the seed crystals were almost same through tests and the surface area of the L-SCMC seeds used was proportional to the amount of seed used. The decreasing rate of concentration of L-SCMC in the solution is considered to be affected by the surface area of seed. And the number of surface nuclei of D-SCMC generated on L-SCMC crystal surface are also supposed to be related to the seed surface area. When the amount of the suspended seed was 4%, the decreasing rate of concentration of L-SCMC was more rapid than that in the solution with 1% of the suspension density, and the induction time for the surface nucleation of D-SCMC on the L-SCMC seed crystal became shorter in spite of same supersaturated concentration of D-SCMC. On the other hand, the surface nucleation is considered to be a probabilistic phenomenon and from these results it is reasonable to assume that the probability of surface nucleation of D-SCMC increases with the surface area of L-SCMC seeds for the range of the suspended seed amount in this study. When surface nuclei of D-SCMC appear more , the decreasing rate of the concentration of D-SCMC in the solution is supposed to be faster, and the results in Figure 2 are reasonably understood .

Table I. Operational conditions

Run.No.	ΔT_{D0}[K]	$\Delta C_{L0} \times 10^2$ [g/100ml]	$\Delta C_{D0} \times 10^2$ [g/100ml]	W of seed [g/100ml]	Suspension density [%]
1-a		4.34	2.84	0.5	0.3
1-b		4.34	2.84	1.7	1.0
1-c	3.6	4.34	2.84	3.4	2.0
1-d		5.34	2.84	3.4	2.0
1-e		5.34	2.84	6.8	4.0
2-a		5.05	3.55	1.7	1.0
2-b	4.6	5.05	3.55	3.4	2.0
2-c		6.05	3.55	6.8	4.0
3-a	5.6	5.74	4.24	1.7	1.0
3-b		5.74	4.24	3.4	2.0
4-a	~	6.39	4.89	0.5	0.3
4-b		6.39	4.89	1.0	0.5
4-c		6.39	4.89	1.7	1.0
4-d		6.89	4.89	1.7	1.0
4-e	6.6	7.39	4.89	1.7	1.0
4-f		7.89	4.89	1.7	1.0
4-g		6.39	4.89	3.4	2.0
4-h		6.39	4.89	6.8	4.0
5-a		7.02	5.52	1.7	1.0
5-b	7.6	7.02	5.52	3.4	2.0
5-c		8.52	5.52	6.8	4.0

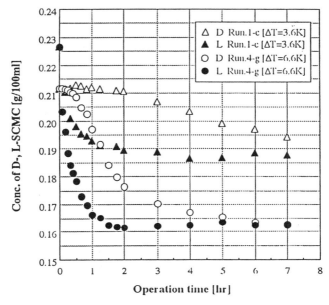

Figure 1. Change of concentration of D-,L-SCMC against operation time [S.D.2%]

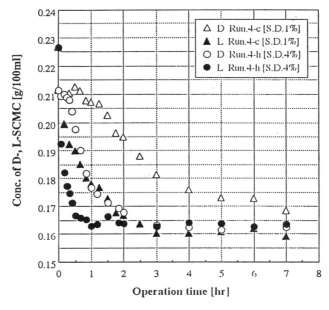

Figure 2. Change of concentration of D-,L-SCMC against operation time [ΔT=6.6K]

Behavior of operational lines expressed by concentration of D-and L-SCMC in the solution.
Operational points expressed by the concentration of D- and L-SCMC in the solution are plotted and operational lines are drawn in Figures 3 and 4. From the results in Figure 3, the saturated concentration is considered to be 0.183g/100ml for D- and L-SCMC. The concentration of L-SCMC alone decreased in the initial stage of operation. When operation continued for some time, the concentration of D-SCMC started to decrease by the growth of surface nuclei of D-SCMC generated on L-SCMC seeds and the operational lines came to bend. These bending points, M in Figure 3 are considered to be decided by the concentration of L-SCMC when the induction time for surface nucleation of D-SCMC passed. The decreasing rate of the concentration of L-SCMC is proportional to the surface area of seeds and to the supersaturation. When the surface nucleation of D-SCMC took place after the concentration of L-SCMC became almost the saturated, operational lines are considered to be bent to parallel to the vertical axis as shown in Figure 3. When the initial supersaturation of D-SCMC is higher, the operational lines are more complicated as shown in Figure 4. After the operational points reached a point M, concentrations of both D- and L-SCMC started to decrease and lines A_1, A_2, A_3 were obtained in tests of suspension density of 0.3, 0.6 and more than 1.0%, respectively. When optical resolution process is considered, operation should be carried out only before the concentration of D-SCMC in the solution starts to decrease. From these plots, operational conditions for optical resolution should be selected on which surface nucleation of D-SCMC does not occur and only L-SCMC crystals grow.

Induction time for surface nucleation of D-SCMC on L-SCMC seed surface.
The induction time required to the D-SCMC nucleation on the L-SCMC seed surface was defined as the elapsed time till the operational lines of D-SCMC started to bend since L-SCMC seeds were fed. This time is plotted against corresponding supersaturation with parameter of suspension density of L-SCMC seeds in Figure 5. From the data obtained from tests for the solution whose supersaturation of D-SCMC was less than 4.89×10^{-2} g/100ml , the induction time was proportional to -2 power of supersaturation. The induction time is also plotted against suspension density in Figure 6 and it is proportional to -0.7 power of suspension density. Then equation 1, is obtained for among the induction time, Θ , initial supersaturation of D-SCMC, ΔC_{D0}, and the suspension density of L-SCMC seed crystal,W.

$$\Theta = 2.43 \times 10^{-3} \Delta C_{D0}^{-2} W^{-0.7} \tag{1}$$

Equation 1, is confirmed by the good correlation of the observed data in Figure 7.
Surface nucleation is generally considered to be a probabilistic phenomenon and the rate of it was studied to be correlated with supersaturation of the solution and the solid surface on which the surface nucleation took place (3). On the other hand, the induction time for nucleation is generally related to nucleation rate. Therefore an induction time required for appearance of D-SCMC nuclei on the surface of L-SCMC seeds is supposed to be affected by the suspension density of seeds and supersaturation of D-SCMC in Figure 5. But the surface nucleation is complicated and from plots in Figure 5, the phenomena in the supersaturated concentration of D-SCMC higher than 5×10^{-2} g/100ml is supposed to be different from that in the solution whose supersaturation of D-SCMC was less than the value. When an optical resolution process is considered, the operation should be selected for the range on which the nucleation of D-SCMC might be neglected and for this restriction, plots in Figure 5 is considered to be useful.

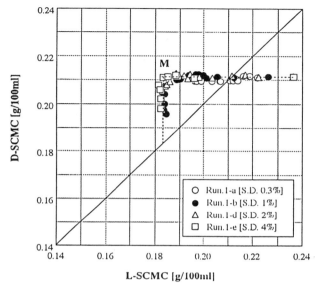

Figure 3. Correlation between the concentration of L-SCMC and that of D-SCMC in the solution since L-SCMC seeds were added in the supersaturated solution   [Δ T=3.6K]

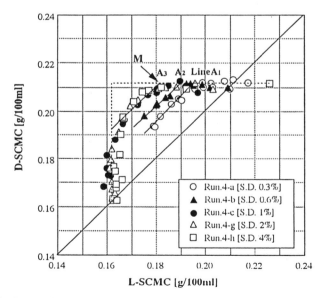

Figure 4. Correlation between the concentration of L-SCMC and that of D-SCMC in the solution since L-SCMC seeds were added in the supersaturated solution   [Δ T=6.6K]

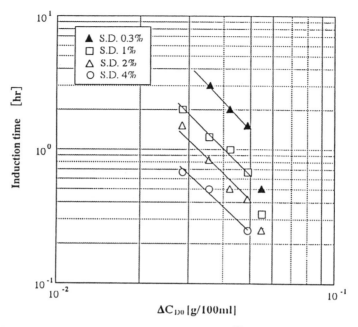

Figure 5. Correlation between induction time Θ and supersaturation of D-SCMC

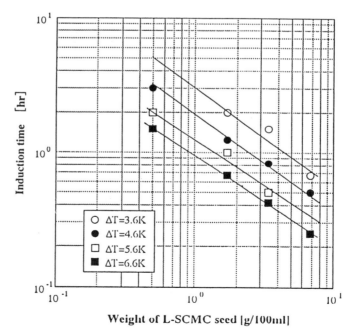

Figure 6. Correlation between induction time Θ and weight of L-SCMC seed

Decreasing rate of concentration of L-SCMC in supersaturated solution of DL-SCMC.
Decreasing rates of concentration of L-SCMC in a solution were calculated from the difference between concentrations of sampled solution obtained at two adjacent sampling times shown in Figure 1,   and are plotted against average supersaturation of the samples in Figure 8 for initial operational supersaturation of D-SCMC Δ C$_{D0}$ of 4.89x10^{-2}g/100ml and the suspension density of 1.0~4.0%.   Plots in this figure are considered to be correlated by two lines whose slopes are different.   The concentration of the solution at the intersection of these two lines corresponds to that of the solution in which both concentrations of D- and L-SCMC are the same.   These lines obtained by different amount of seed are parallel to one another.   The decrease of the concentration of L-SCMC in a solution is considered to depend on the growth of the L-SCMC seeds suspended in the supersaturated solution, so the decreasing rate of the concentration is supposed to be correlated with equation 2.

$$-V\frac{dC_L}{d\theta} = k_L(W)^m(C_L - C_{LS})^n \qquad (2)$$

Here, k$_L$, W, and V are mass transfer coefficient of L-SCMC, amount of L-SCMC seed, and volume of the solution, respectively, and m and n are power numbers of W and supersaturation .   The amount of L-SCMC grown in a batch operation in these tests was less than one percent of the amount of seed crystal.   Therefore, k$_L$ (W)^m/V is assumed to be constant and from plots in Figure 8, equations 3, and 4, are obtained.

For the range in which the concentration of L-SCMC is larger than that of D-SCMC

$$-V\frac{dC_L}{d\theta} = 234W(C_L - C_{LS})^{2.7} \qquad (3)$$

For the range in which the concentration of L-SCMC is less than that of D-SCMC

$$-V\frac{dC_L}{d\theta} = 0.78W(C_L - C_{LS})^{1.0} \qquad (4)$$

Data obtained from Runs.1 to 5 in Table. I are also plotted in Figure 9 and 10, for the range of C$_L \geq$C$_D$ and C$_L \leq$C$_D$, respectively.   From these figures, equations 3, and 4, are considered to show good correlation between growth rate, amount of seed crystal and supersaturation.   Crystal surface area is supposed to be almost proportional to the amount of suspended crystal and crystal volume is considered to be mainly affected by longitudinal length by the reason of almost same lateral area. The connecting points between the two lines particularized by C$_L \geq$C$_D$ and C$_L \leq$C$_D$ is considered to be correspond to the supersaturation of L-SCMC which is same the initial supersaturation of D-SCMC.

Application of the correlations of induction time of surface nucleation and decreasing rate of the concentration of the solution to optical resolution process of DL-SCMC by crystallization.
Optical resolution of DL-SCMC by crystallization is carried out by growth of D-SCMC and L-SCMC seed crystal in DL-SCMC supersaturated solution, in a respectively separated crystallizer for growth of either crystal.   The skeleton of the main parts of the process is shown in Figure 11. The volume of the crystallizer is

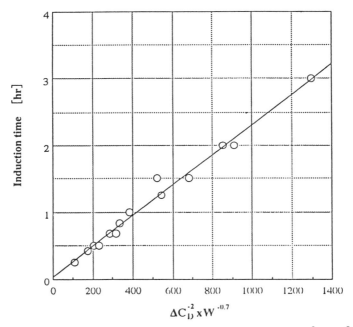

Figure 7. Correlation between induction time Θ and $\Delta C_D^{-2} x W^{-0.7}$

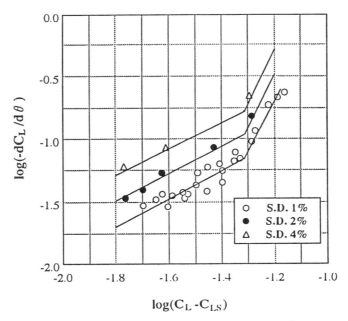

Figure 8. Change of log(-dC_L/d θ) against log(C_L-C_{LS}) [Δ T=6.6K]

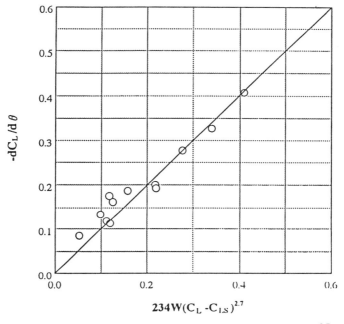

Figure 9. Correlation between $-dC_L/d\theta$ and $234W(C_L-C_{LS})^{2.7}$

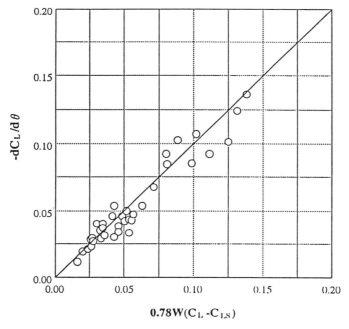

Figure 10. Correlation between $-dC_L/d\theta$ and $0.78W(C_L-C_{LS})$

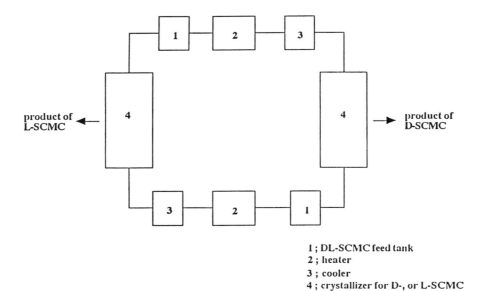

1 ; DL-SCMC feed tank
2 ; heater
3 ; cooler
4 ; crystallizer for D-, or L-SCMC

Figure 11. Skeleton of optical resolution process by crystallization

estimated using correlations obtained in this study.   In general, it is difficult to crystallize DL-SCMC in an aqueous solution. When sodium chloride is dissolved in it, DL-SCMC crystal easily come out from supersaturated solution and L-SCMC seed crystals also grow without crystallization of D-SCMC under restricted operational conditions (4).   The correlative equations used for estimation of volume of a crystallizer are obtained as follows.

The decreasing rate of concentration of L-SCMC is correlated by equations 3, and 4, respectively for $C_L \geq C_D$ and $C_L \leq C_D$, and equations 5, and 6, are derived from the amount of product crystal of L-SCMC for either range of $C_L \geq C_D$ or $C_L \leq C_D$, respectively.

For $C_L \geq C_D$

$$P_1 = \int_{C_R}^{C_{L0}} V dC_L = \int_0^{\theta_1} 234W(C_L - C_{LS})^{2.7} d\theta \tag{5}$$

For $C_L \leq C_D$

$$P_2 = \int_{2C_R - C_{L0}}^{C_R} V dC_L = \int_0^{\theta_2} 0.78W(C_L - C_{LS})^{1.0} d\theta \tag{6}$$

Operational time for one batch, θ^*, in which concentration of L-SCMC decreases from C_{L0} to $2C_R - C_{L0}$ is given by equation 7.

$$\theta^* = \theta_1 + \theta_2 \tag{7}$$

θ_1 and θ_2 are calculated from equations 8, and 9, derived from equations 3, and 4.

$$\theta_1 = \frac{V}{1.7 \times 234W}\left[\frac{1}{(C_R - C_{LS})^{1.7}} - \frac{1}{(C_{L0} - C_{LS})^{1.7}}\right] \tag{8}$$

$$\theta_2 = \frac{V}{0.78W} In\left(\frac{C_R - C_{LS}}{2C_R - C_{L0} - C_{LS}}\right) \tag{9}$$

On the other hand, the amount of L-SCMC crystals produced in a batch operation, P^*, is calculated by equation 10,

$$P^* = P_1 + P_2 \tag{10}$$

Here, P_1 and P_2 are obtained from equations 5, and 6, respectively.

Equations 5, to 10, are used for estimation of the volume of a crystallizer to produce L-SCMC or D-SCMC of P kg/hr as follows.

In this case, the production rate of D-SCMC and L-SCMC is considered to be the same and expressed by P^*.   The concentration of DL-SCMC of feed is set by $2C_R$. Operational temperature is assumed to be set and then the saturated concentration of D-SCMC and L-SCMC in the solution is expressed by C_{DS} and C_{LS} decided by the operational temperature.   In steady state operation, the concentration of D- or L-

SCMC in feed and outlet solution are considered to be determined by C_{D0} or C_{L0} and $2C_R-C_{D0}$ or $2C_R-C_{L0}$, respectively.

In order to estimate the volume of a crystallizer, V, W/V and C_{D0} or C_{L0} should be assumed at first, and then the volume of a crystallizer is estimated as follows (Example is shown for L-SCMC). From equations 8, and 9, operational time for L-SCMC, θ_1 and θ_2 are easily estimated. Then total operational time, θ^*, for one batch is obtained from equation 7. The amount of product crystal of L-SCMC P_1 or P_2 is also estimated for assumed volume of V_L from equation 5, or 6, for either case of $C_L \geqq C_D$ or $C_L \leqq C_D$. Then total product amount of P^* for one batch is also obtained by equation 10. On the other hand, desired production rate of L-SCMC,P^*, should be equal to (P^*/θ^*) and then volume of crystallizer for L-SCMC,V_L, is easily decided. In this process, θ^* should be less than the induction time for surface nucleation of D-SCMC, Θ, estimated from equation 1. Therefore, assumed W/V and C_{L0} should be finally decided from these restrictions.

In order to apply these calculation results to industrial purposes, some additional studies on treatment for prevention of the surface nucleation of D-SCMC is considered to be required.

Conclusions.

Concentrations of D-SCMC and L-SCMC in an agitated supersaturated solution were observed on operational conditions , volume of solution 500ml, suspension density of L-SCMC seed 0.3~4.0% for initial supersaturation of L-SCMC seed 4.34×10^{-2} ~8.52×10^{-2} g/100ml, for initial supersaturation of D-SCMC 2.84×10^{-2}~5.52×10^{-2} g/100ml, and the following results were obtained.

1. The induction time required for surface nucleation of D-SCMC crystal is correlated with supersaturation of D-SCMC and suspension density of L-SCMC seeds.

2. The decreasing rate of concentration of L-SCMC in DL-SCMC supersaturated solution was observed and correlative equations 3, and 4, were obtained.

For $C_L \geqq C_D$

$$-V\frac{dC_L}{d\theta} = 234W(C_L - C_{LS})^{2.7} \tag{3}$$

For $C_L \leqq C_D$

$$-V\frac{dC_L}{d\theta} = 0.78W(C_L - C_{LS})^{1.0} \tag{4}$$

3. A decision of volume of a crystallizer for optical resolution process was proposed and discussed.

Literature Cited.

(1) Yokota, M.; Toyokura, K. *Ind. Crystallization'93.* **1993**, *vol.2*, pp.*203-208*.
(2) Yokota, M.; Toyokura, K. *Chemical Engineering.* **1994**, *vol.39*, pp.*69-76*.
(3) Turnbull, D.; Vonnegut, B. Ind. Eng. Chem. **1952**, *vol.44*, pp.*1292*.
(4) Yokota, M.; Toyokura, K. J. of Crystal Growth 99. **1990**, pp.*1112-1116*.

Chapter 6

Purity Drop in Optical Resolution of Racemic Mixtures

M. Matsuoka

Department of Chemical Engineering, Tokyo University of Agriculture and Technology, 24–16 Nakacho-2, Koganei, Tokyo 184, Japan

Phenomena of purity drop in preferential crystallization for resolution of racemic mixtures are first surveyed, and proposed mechanisms relevant to the purity drop are then summarized to understand properly the phenomena. Finally with the aid of additional experimental data a possible mechanism of such spontaneous crystallization resulting in the purity drop is discussed and a process to prevent purity drop is also proposed.

Optical resolution by preferential crystallization has long been studied to recover or to separate a desired enantiomer (optically active isomer) from a conglomerate (racemic mixture) solution. By preferential crystallization we refer to separation methods by crystallization of a desired isomer from its conglomerate by seeding crystals of the desired isomer. It is essential for given systems to form conglomerates, i.e. not to form either solid solutions or racemic compounds. Conglomerates are heterogeneous mixtures of both enantiomers at equal amounts. A number of systems feasible to resolve by preferential crystallization have been found on the basis of the solid-liquid phase equilibria (1).

Practically preferential crystallizations are carried out as batchwise operations, and it has been pointed out that an undesired enantiomer starts to crystallize at some later stages of the operation so that a purity drop of the product crystals results. Figure 1 shows an example of such optical purity drop phenomena (2): after four hours of resolution the optical purity suddenly started to decrease, where the solution concentration is defined as the pseudo binary mixture excluding the solvent, i.e. $100 \times$ L-isomer/(L-isomer+D-isomer) when the L-enantiomer is the desired isomer. The optical purity (op) is related to the conventional purity (x in mol fraction of L-isomer) by the equation of $op = 100 \times (2x-1)$. During the resolution it can be seen that the concentration of the desired enantiomer in the solution decreased and after four hours the decrease ceased and then the concentration started to gradually increase again to approach 50% when the concentration is defined as the fraction of the desired isomer in the total isomer contents in the solution.

In order to clarify the mechanism of the purity drop phenomena it is essential to consider how and why the unseeded (undesired) enantiomer crystals start to appear;

without such appearance in terms of nucleation of this enantiomer the product purity would be kept high until the final piece of crystals of the desired enantiomer deposits.

Since comprehensive knowledge of phase equilibria, crystallization phenomena, crystallization kinetics and process controls is required to establish a process to produce optically high purity materials with high yields,   preferential crystallization is undoubtedly a challenging topic for those working in the field of industrial crystallization. In this article, the relation between the spontaneous nucleation and the phase equilibrium will be first discussed.   A brief survey of spontaneous nucleation phenomena will follow.   Then our experimental work on the effect of pretreatment of seed crystals will be discussed.

Fundamentals of Purity Drop

Principles of Preferential Crystallization on Phase Diagrams.   Solutions containing racemic mixtures include a solvent and hence are at least ternary systems. A representative ternary solubility curve (solid liquid phase equilibrium at constant pressure and temperature) is schematically illustrated in Figure 2 showing immiscibility between the isomers and that the supersaturated solution of the conglomerate (indicated as point M) lies in the area where both isomers exist as the crystal phase (solid D and solid L) and coexist with the saturated solution of which composition is given by point E in Figure 2.   During the progress of resolution, i.e. crystallization of one isomer with the other one remaining in the solution, the concentration of the crystallizing component decrease as already shown in Fig.1, while that of the other isomer will relatively increase.   However depending on the phase diagram the supersaturation of the unseeded isomer will increase or decrease during resolution.

The critical situation occurs when the solubility ratio α , defined as the ratio S_R/S_D where S_R and S_D denote the solubilities of the conglomerate and D-isomer respectively, is equal to 2, which is typical for slightly dissociated materials (1). Under such conditions crystallization of the seeded isomer does not change the supersaturation of the unseeded isomer, so that spontaneous crystallization or more exactly spontaneous nucleation would not be expected.   On the other hand, systems having α less than 2 would be expected more stable.   This will be discussed in more detail in a coming section.

Crystallization and Mass Balance.   For optical resolution by preferential crystalliza-tion of racemic mixtures two enantiomers exist in the feed solution at an equal amount as the initial mixture, the supersaturations of the both components being the same.   During the resolution the supersaturation of the desired or seeded enantiomer decreases due to the deposition onto the seed crystals and approaches to a certain composition (point A in Figure 3) given as the intersecting point of the line connecting the feed (M) and the pure desired component (L) and the extrapolated solubility curve of the desired enantiomer (S_L-S_L'). Meanwhile the supersaturation of the unseeded enantiomer would either increase, remain constant or decrease depending on the solubility curves of the system.   Accordingly the possibility of primary nucleation of the other enantiomer is expected to increase, remain constant or decrease respectively.

Supersaturated racemic solutions are unstable or metastable depending on the degree of supersaturation. For a system having a phase diagram as shown in Figure 3, the initial levels of the driving force for the crystallization are the same to the both enantiomers, i.e. AM for the L-isomer and BM for the D-isomer.   Accordingly at the end of the complete crystallization when the true equilibrium is established,   i.e. when no resolution results, the saturated solution (E) and two crystal phases of the enantiomers (L and D) of the equal amounts coexist.   There may be no need to mention that the problem of the purity drop is attributed to the crystallization of the undesired enantiomer, often it is called spontaneous crystallization and that the

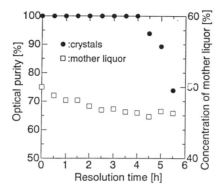

Fig.1   An example of purity drop phenomena in preferential crystallization of L-threonine (*L-Thr*)

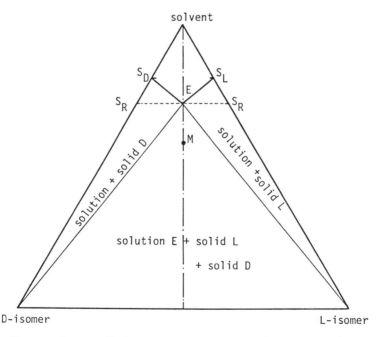

Fig.2   Ternary phase equilibrium at constant pressure and temperature for racemic mixture (conglomerate) systems

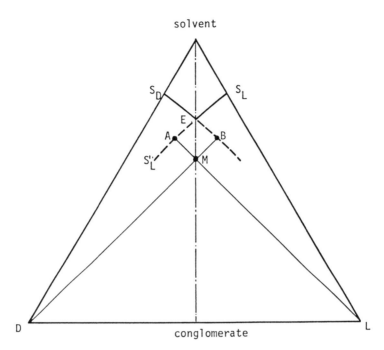

Fig.3   Mass balance and equilibrium during resolution

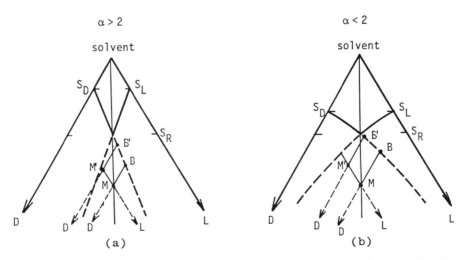

Fig.4   Solubility curves with the solubility ratio greater than 2 and those less than 2

problem is how to prevent the *spontaneous* crystallization of the unseeded enantiomer during the resolution .

Occurrence of Spontaneous Crystallization of an Unseeded Enantiomer.   When we consider the purity drop on the phase diagram, it is the prerequisite that the seeds added to cause the preferential resolution are completely pure.   However if the seeds are contaminated by the undesired enantiomer, the phenomena could be completely different.   Since in general seeds are taken from a previous batch, they are likely to have been contaminated by the adhered mother liquor due to incomplete phase separation or by some non-equilibrium crystallization phenomena such as inclusion of mother liquor or agglomeration of crystalline particles.   For the case of the former, drying of the crystals generates small particles of the undesired enantiomer which will then be introduced to a racemic solution as seeds.

Factors and Phenomena of Purity Drop: Survey of Previous Studies

The key factors influencing the purity drop phenomena are considered at first.   They must be the factors which enhance primary nucleation or growth of attached tiny crystalline particles on the seed crystals; therefore a list of factors would include: solution supercooling (or supersaturation), agitation speed, suspension density, mass of seed crystals, pretreatment of seed crystals, cooling rate and phase equilibria of given systems.   Effects of these factors investigated in literature are examined separately.

Phase Equilibria.   As mentioned in the preceding sections, the solid-liquid phase equilibrium of a given system is considered to be essential for spontaneous nucleation to occur. There are two ideal cases for the phase equilibria of conglomerates forming systems.   For enantiomers which are not dissociated in the solution phase, the ideal solubility of the conglomerate is two times the individual enantiomers, i.e. $\alpha = 2$ where the solubility ratio α is defines as mentioned earlier as the ratio of the solubility of the conglomerate to the solubility of the individual enantiomer $\alpha \equiv S_R/S_L$, while when the enantiomers are highly dissociated $\alpha = \sqrt{2}$ (*1*).   For systems having low solubilities $(S_R, S_L < 1)$ in which the changes in the composition during the resolution can be approximated as a shift on a line parallel with the triangle base, the value of α can be a criterion to judge whether the supersaturation of the unseeded enantiomer would increase or decrease during the resolution.   Since the solubility curves are parallel to the bases for systems with $\alpha = 2$, systems having α greater than 2 are unstable because of the fact the MB < M'B' as shown in Figure 4-a.   On the other hand those systems having α less than 2 are expected to be stable, since MB > M'B' (Figure 4-b) so spontaneous nucleation would not occur.   If the spontaneous crystallization of the unseeded enantiomer is caused by primary nucleation the phase equilibrium would thus be essential. Jacques et al. (*1*) give a list of the solubility ratios for many organic systems.
    In addition to the solubility behaviour, the selection of solvents is also important since solvents can change the solubilities of the enantiomers under consideration. Shiraiwa et al. (*3*) reported that the control of the solvent composition could eliminate the purity drop by the preferential crystallization for the racemic compound forming system of (*RS*)-Tyrosine 4-chlorobenzenesulfonate (Figure 5) in which the solvents examined were mixtures of ethanol and 4-chlorobenzene sulfonic acid (CBA).

Degree of Supercooling.   There are reports on the existence of metastable zones for many systems.   Hongo et al. (*4*) discussed the existence of the first metastability region (SI) for the aqueous DL-serine m-xylene-4-sulphonate dihydrate (DL-Ser·mXS·2H$_2$O) system which was defined as the region where no spontaneous crystallization occurred for about 4 hours or more (Figure 6).   They mentioned that the stability limits were independent of the cooling rate of the solutions.   They further

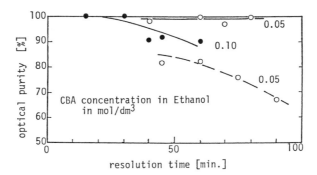

Fig.5   Effects of solvent compositions on purity drop for the RS-Tyr 4-CBA system

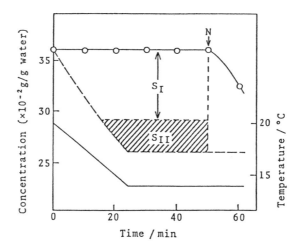

Fig.6   Definitions of metastability limits given by Hongo et al.
(Reproduced with permission from ref.4. Copyright 1981 The Japanese
Chemical Society)

discussed the second metastable region (SII) which was defined as the integrated area with respect to the time and the supersaturation during the resolution. The second metastable region was not affected by the cooling rate nor the lowest arrival temperature.

Ohtsuki (5) summarized their experimental data such as the critical supersaturation above which spontaneous crystallization of the undesired enantiomer occurred as a function of solution temperature for the DL-Threonine - water system (Figure 7). The critical supersaturation was defined there as the supercooling limits where no spontaneous crystallization of the unseeded enantiomer was observed in two hours. We (6) have found that the purity drop started soon after the resolution when the initial supercooling was as high as 10K, while no purity drop was observed in 3 hours when it was 5°C as shown in Figure 8. Yokota et al. (7) also mentioned the effects of the supercooling on the product purity for S-carboxymethyl-D-cysteine (D-SMSC); the purity was lower at higher supercoolings.

From these observations, the following conclusion could be deduced. There exist supercooling for above which spontaneous crystallization of unseeded (undesired) enantiomers occurs, however purity drop could hardly be prevented if the solution is kept at supersaturated conditions for long periods of time.

Agitation Speed. Without theoretical explanations agitation is known to enhance the purity drop phenomena or decrease the product purity from the preferential crystallization. We (6) have shown that the product purity drop is decreased by increasing the agitation of the solutions (Figure 9). The increase of the agitation speed from 200 to 400 rpm reduced the purity from about 95 to 80%. Similar tendency was reported by Yokota et al.(7) for SCMC crystallization.

Suspension Density. The concentration of crystals in the solution could decrease the product purity of threonine as shown in Figure 10 when the concentration exceeded 0.1 g/ml for almost constant supersaturations for threonine (5). The effect is very clear, although the reason is not understood. The amount of seed crystals has similar influence on the start of the purity drop for the threonine system (6). A large amount of seed crystals does shorten the time at which the purity drop starts.

Washing of seed crystals. Washing the seed crystals could delay the start of purity drop (6). From the observation with SEM the surfaces of seed crystals were very much different before and after washing with water; before washing the surfaces were covered with a number of small crystalline particles, while they were very smooth after washing due to partial dissolution.

Effects of Seed Pretreatment : Rate of Crystallization and the Number of Recrystallization

Since the above mentioned experimental evidences suggest primarily that the major causes of the purity drop may come from the seed crystals added to the racemic solutions, some additional experiments were conducted to examine the effects of the quality of seed crystals (Takiyama,H. et al., TUAT, unpublished data). First, the rate of crystallization, i.e. the cooling rate of the solution, and the holding time at the final temperature after the crystallization were experimentally examined. In another series of experiments in order to eliminate any possible undesired enantiomer particles from the seed crystals, seeds are recrystallized for a number of times.

The effects of the preparation methods and the number of recrystallization were examined experimentally by the use of the threonine (*Thr*.) - water system. The supersaturation of *D-Thr*., for example, will slightly increase during the crystallization of *L-Thr* since for this system $\alpha < 2$. This may lead to a conclusion that the possibility of primary nucleation to occur would also slightly increase as the progress of the resolution. Small stirred tank crystallizers were used both for preparation of

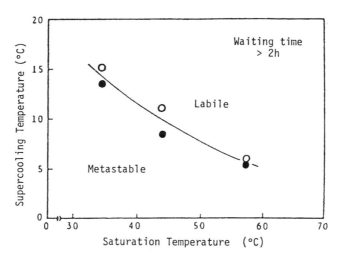

Fig.7 Metastable region for threonine
(Reproduced with permission from ref.5. Copyright 1982 The Society of
Chemical Engineers, Japan)

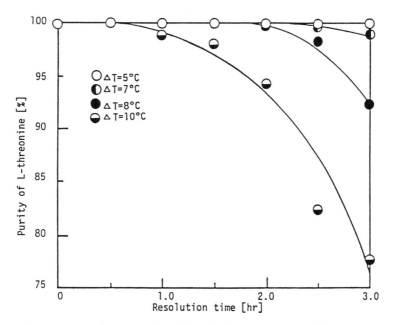

Fig.8 Changes of product optical purities with time for different initial supercoolings

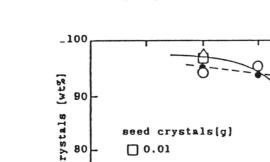

Fig.9   Effect of agitation speed on product purity for threonine

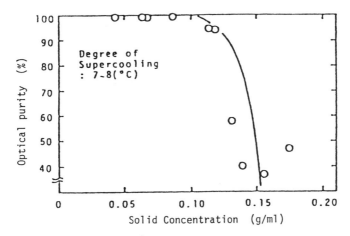

Fig.10   Effects of suspension density on product purity
(Reproduced with permission from ref.5. Copyright 1982 The Society of
Chemical Engineers, Japan)

seed crystals and for resolution experiments. The purity of the crystals and the solution concentration were analyzed by HPLC. Care has been taken to eliminate any possible contamination by the mother liquor during sampling and phase separation.

Results.   The following results were obtained for the effects of preparation methods of the seeds.
1. Seeds prepared by fast cooling improved the purity drop.
2. Seeds held for a longer period in the crystallizer with agitation after crystallization caused purity drop of the resulting crystals to occur easily.

These facts suggests that for better separation seed crystals should be prepared quickly.   The size distribution of crystals kept in a crystallizer for long time showed increased numbers of smaller particles, hence larger surface areas.

Under adequate conditions determined for the preparation of the seeds, recrystallization was repeated and the obtained crystals were used as seeds.   As the number of recrystallization (N) increases the time of the start of purity drop was prolonged, implying that recrystallization is effective in retarding the nucleation of L-Thr.   The effect, however, seems to have limits since there is little improvement in the starting time of purity drop above, say $N=5$, which is not shown in Figure 11 to avoid confusion.

In addition purity of fine particles was found always lower in comparison with those of larger or middle sized particles as shown in Figure 12. This suggests that the particles generated in the later stages are mixtures of primary and secondary nuclei, where the primary nuclei would be of the unseeded enantiomer while the secondary nuclei the mixtures of seeded and unseeded enantiomers.

Discussion

Summary of Experimental Evidence.   The above mentioned facts are summarized in Table 1.   From the facts obtained in our studies we can consider the mechanism of the purity drop as one that a very small amount of *D-Thr.* present on the surface of the seeds or inside of the seeds which happen to be exposed on breakage starts to grow in the supersaturated solution until some critical size when breeding (secondary nucleation) bursts in the solution.   The time needed for *D-Thr.* to grow to the critical size and the number of such elementary *D-Thr.* determine the time of the start of purity drop.

In addition to this secondary nucleation phenomena, we should consider other possibilities to explain the reason why the recrystallization did not completely eliminate spontaneous crystallization.   If the purity drop is attributed to *D-Thr* in the seeds, repetition of recrystallization could completely prevent from nucleation of the undesired enantiomer.   However this is not the case.   *L-Thr.* crystals somehow nucleate after a few hours of crystallization of *D-Thr.*

Then the possibility of primary nucleation should be considered.   Or we might assume two dimensional nucleation of *D-Thr* on the surface of the *L-Thr* substrate crystals as proposed by Yokota et al. for the D-SCMC system.

Instead we should discuss the possibility of primary nucleation for the system. For racemic solutions of threonine as mentioned above the deposition of *L-Thr.* results in the increase of the supersaturation of *L-Thr.* therefore the possibility of primary nucleation will increase.   For the SCMC system, unfortunately the solubility data are not available in their paper so further discussion cannot be made at the moment.

Interestingly Davey et al. (8) reported the phenomena that a single large piece of triazolylketone crystal prepared from its racemic solution was racemate although it looked like a single crystal having a hour glass shaped shadow in it.   They cut it into 27 fragments and analyzed chiralities of each piece to find that they were widely dispersed.   Their conclusion is therefore "that growing surfaces of triazolylketone crystals can accommodate, with little change in interfacial tension and binding energy,

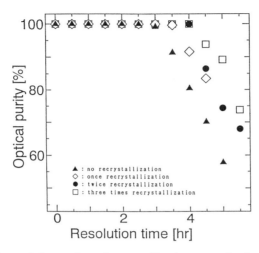

Fig.11 Effects of the number of recrystallization on purity drop occurrence

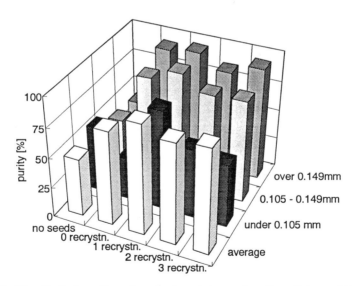

Fig.12 Optical purities of product crystals as a function of size and time

Table 1   Factors of and their effects on purity drop in preferential crystallization

factor	subfactor	effect[*]	reference
phase equilibrium	solubility ratio	major	Jacques et al. (*1*)
	solvent selection	major	Shiraiwa (*3*)
operating condition	supercooling	critical	Hongo, et al. (*4*)
			Ohtsuki (*5*)
			Matsuoka et al. (*6*)
	cooling rate	minor	Hongo et al.(*4*)
	agitation	considerable	Matsuoka et al. (*6*)
		minor	Yokota et al. (*7*)
	suspension density	major	Ohtsuki (*5*)
seed crystal	preparation method	considerable	Present authors
	amount	considerable	Matsuoka et al. (*6*)
	washing	critical	Matsuoka et al. (*6*)
	recrystallization	major	Present authors

[*]   The order of the effect is:
      critical > major > considerable > minor.
      The effects can be to enhance or to reduce the purity drop phenomena.

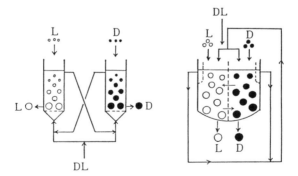

Fig.13   Illustrative processes to prevent from spontaneous crystallization of
            undesired D-isomer in product flows of L-isomer
            (Reproduced with permission from ref.9. Copyright 1983
             The Society of Chemical Engineers, Japan)

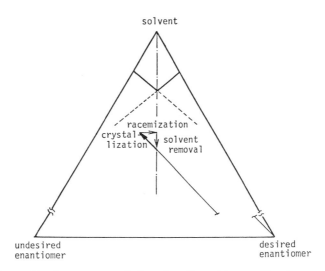

Fig.14   A combined process consisting of preferential crystallization, racemization and solvent evaporation

molecules of either chirality". Furthermore they mentioned that "this might lead either to a growing surface on which islands (nuclei) of both chirality are present or to a coherent layer of an interfacial racemic compound onto which the opposite isomer can grow". This may suggest the cogrowth of L-enantiomer on D-seeds, but we need further study on such phenomenon for racemic mixture systems.

For industrial application we can conclude that the best and simple way to obtain a pure enantiomer from its racemic mixture by preferential crystallization would be to use washed seed crystals.

Proposal of A Process To Prevent Purity Drop

In order to prevent spontaneous crystallization of the other isomers, processes to remove concentration (supersaturation) of the other isomers are essential. In general this has been done by crystallization of the undesired enantiomer in parallel or in series with the crystallization of the desired enantiomer. Figure 13 (9) illustrates some parallel processes where undesired enantiomers are crystallized either in a separate crystallizer where mother liquors circulate or in a single crystallizer with separated space. The obtained undesired enantiomer is then converted into the desired enantiomer by the racemization reaction to improve the yield of the desired component. Our proposal is to combine the preferential crystallization of the desired isomer with the racemization reaction in a single crystallization vessel (*10*). The idea is not new and is outlined in a book by Jacques et al. where the use of aldehyde or ketone as the catalyst for the racemization reaction is suggested.

Experiments were carried out using the D-/L-DOPA (β-(3,4-dihydroxyphenyl) alanine) system with organic acids as the solvents and organic aldehydes as the catalysts for the racemization reaction. Under adequate conditions empirically determined, three rate processes of preferential crystallization, racemization reaction and solvent evaporation proceeded simultaneously at the same rate and the solution composition would then be kept unvaried at the same one as that of initial mixture as shown Figure 14. This process is therefore possible to operate continuously by feeding a conglomerate mixture and by removing the product crystals and the solvents.

There would be no accumulation of the undesired isomer in the solution.   Any excess amounts of the undesired enantiomer in the solution could be the driving force for the isomerization reaction.

This process is promising in that the yield could be twice since the undesired enantiomer is converted into the desired one and the product purity would be maintained high enough particularly when continuous operations are introduced. This process is currently under study.

Conclusions

Phenomena of purity drop in racemic resolution by preferential crystallization were surveyed and factors relevant to spontaneous crystallization and their significance were quantitatively summarized.   It was empirically concluded that washing of seed crystals would be the best way to prevent such purity drop, although it could not completely suppress the occurrence.   With additional experiments on the preparation method of seed crystals, adequate procedures were found.   However, recrystallization under these conditions could only prolong the start of purity drop but could not prevent it.   A combined process to produce optically pure enantiomers consisting of preferential crystallization, racemization reaction and solvent evaporation steps was proposed.

Acknowledgments

The author is grateful to Dr. H.Takiyama and K.Masumiya for their discussion in preparing the present article.

Literature Cited

1.   Jacques,J.; A. Collet,A. Wilen,S.H. "Enantiomers, Racemates and Resolutions", John-Wiley & Sons, Inc. New York, 1981.
2.   Sugawara,S.; Matsuoka,M. Preprints of the 26th Autumn Meeting of the Soc. Chem. Eng., Japan, 1993.
3.   Shiraiwa,T.; Sumami,M.; Kakimoto,M.; Inui,S.; Kurokawa,H. *Technol. Rep. Kansai Univ.*, **1991**, 33, 113.
4.   Hongo, C.; Yamada, S.; I.Chibata,I. *Bull. Chem. Soc., Jpn.*, **1981**, 54, 1905 and 1911.
5.   Ohtsuki,M. *Kagaku Kikai Gijutsu* (*Technology of Chemical Machinary*)34, The Society of Chemical Engineers, Japan,1982, p.31.
6.   Matsuoka,M; Hasegawa,H.; Ohori,K. Crystallization as a Separation Process", ACS Ser. 438, 1990, Chap.18.
7.   Yokota,M.; Oguchi,T.; Arai,K.; Toyokura,K.; Inoue,C.; Naijo,H. Crystallization as a Separation Process, ACS Ser. 438, 1990, Chap.20.
8.   Davey,R.J.; Black,S.N.; Williams,L.J.; McEwan,D; Sadler,D.E. *J. Crystal Growth*, **1990**, 102, 97.
9.   Kitahara,T., *Kagaku Kogaku*, **1983**, 47, 227.
10.   M.Matsuoka, H.Hasegawa and K.Nakase, 1992, Preprints of the 25th Autumn Meeting of the Soc. Chem. Eng., Japan.

Chapter 7

Separation of L-Mandelic Acid from Asymmetric Mixtures by Means of High-Pressure Crystallization

N. Nishiguchi[1,3], M. Moritoki[1,4], T. Shinohara[2], and Ken Toyokura[2]

**[1]Engineering and Machinery Division, Kobe Steel, Ltd.,
3–18 Wakinohamacho 1-Chome, Chuoku, Kobe 651, Japan
[2]Department of Applied Chemistry, Waseda University, 3–4–1 Okubo,
Shinjuku-ku, Tokyo 169, Japan**

A method using high pressure crystallization was studied for separating L-mandelic acid from asymmetric mixtures. The solubility of various mixtures of D- and L-mandelic acid in water was measured under high pressure. It was found that the solubility decreases with increase in pressure, and that the L/D composition of the eutectic compounds formed between a chiral (D-, or L-form) and racemic compound (L/D=50/50) shifts, respectively, toward the racemic composition with increasing pressure. It was also found that a high purity chiral is obtainable using high pressure crystallization from asymmetric mixtures of D- and L-form, whose concentrations are in the region of the racemic compound formation at atmospheric pressure, and that the yield almost agrees with those calculated from the solubility curves. A novel process was proposed for L-mandelic acid separation.

Chiral (L-, or D-form) separation from an optical isomer is essential for its commercial use. Preferential crystallization is often used as a separation technique of a chiral from optical isomers, but it is invalid when racemic compounds are formed, as it is in the production of mandelic acid by chemical reaction. However, by using high pressure crystallization, it becomes possible to separate a chiral from the mixture in the region of racemic compound formation under atmospheric pressure, when the eutectic compositions shift toward the racemic composition under high pressure.

Principle of Separation

The solubility curves of mandelic acid have eutectic points between the chiral and the racemic compound. At atmospheric pressure, the eutectic composition is almost constant regardless of temperature, while the solubility varies according to the temperature. So, it is hard to separate the chiral crystal from the feed when the composition is between the eutectic and the racemic composition.

[3]Current address: Engineering and Machinery Division, Kobe Steel, Ltd., 6–14 Edobori 1-Chome, Nishi-ku, Osaka 550, Japan

[4]Current address: 11–6 Higashi 3-Chome, Midorigaoka-cho, Mikishi, Hyogo 673–05, Japan

However, the region of chiral formation comes to be wider under high pressure, because the eutectic point shifts toward the racemic composition.

In this case, chiral crystal can be separated by combining cooling crystallization and high pressure crystallization. The solution mixture with a nearly eutectic composition at atmospheric pressure can be obtained with concentration by evaporation and by cooling crystallization of racemic compounds (Figure 1, Step 3). Chiral of high purity can be separated from this mixture by high pressure crystallization (Figure 1, Step 1 and Step 2).

In this research, a possibility of separating the chiral from D-, L-mandelic acid solutions was studied on the basis of the idea shown in Figure 2. At first, the solubility of mandelic acid in water was measured under high pressure, and then the experiments separating L-mandelic acid from its D- and L-mixtures in Zone I and in Zone II were performed by use of high pressure crystallization. Finally, a novel separation process is proposed for mandelic acid.

Experiment

Measurement of solubility   The outline of the apparatus used in this study is shown in Figure 3 (*1*). A feed mixture of a given composition was put into the high pressure vessel in an oil bath kept at 303K and was pressurized stepwise to a given separating pressure, which was then held constant. The liquid-solid equilibrium was estimated to be attained within 2 hours, because the piston displacement became constant. Thus, the pure water in the reservoir vessel was pressurized beforehand to the same pressure in order to avoid the pressure fall at the beginning of transferring the liquid. Then, the valve V1 was opened to transfer the mother liquor from the solid in the crystallization vessel. In this procedure, a little pulsation of liquid pressure could not be avoided. The mother liquor which was transferred from the crystallization vessel to the reservoir vessel was removed through the valve V2 and was used for analysis. The concentration of D-, and L-mandelic acid in this mother liquor was analyzed by High Performance Liquid Chromatography (HPLC) and the solubility was determined from the average concentration. The operational pressures in these experiment were 176.4, 215.6, 254.8, and 294.0 MPa. The feeds used were the four samples shown below.

Feed A : 25g mandelic acid ╱100g water (L:D = 90:10)
Feed B : 30g mandelic acid ╱100g water (L:D = 85:15)
Feed C : 60g mandelic acid ╱100g water (L:D = 70:30)
Feed D : 25g mandelic acid ╱100g water (L = 100)

Separation of L-form in Zone I and in Zone II   The same experimental apparatus as mentioned above was used. The operating method and the conditions were also the same as described above. In Zone I, the feeds used in these experiments were as follows.

Feed A : 25g mandelic acid ╱100g water (L:D = 90:10)
Feed B : 30g mandelic acid ╱100g water (L:D = 85:15)

The amount of L-form in these feeds was controlled to be higher than the eutectic solubility of mandelic acid and also the amount of D-form in these was controlled to be a value lower than the eutectic solubility at a given pressure. The weight and composition of D-form (D/L+D) and L-form (L/L+D) of the crystal obtained in this experiment were measured with HPLC method. In Zone II, Feed C was used.

Feed C : 60g mandelic acid ╱100g water (L:D =70:30)

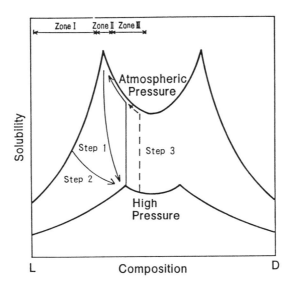

Figure 1.   Principle of L-form separation

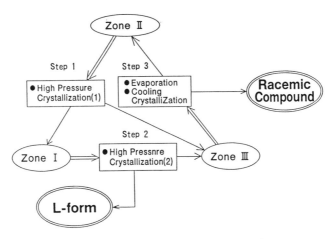

Figure 2.   Concept of L-form separation

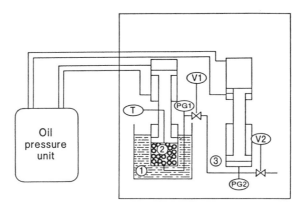

1. Oil bath   T.   Thermocouple
2. High pressure vessel   V.   Valve
   for crystallization   PG. Pressure gauge
3. High pressure
   reservoir vessel

Figure 3.   Schematic diagram of experimental apparatus

Experimental results

Results of solubility measurement   The solubility curves of D-,L-mandelic acid at atmospheric pressure were reported previously (2) as shown in Figure 4 for temperature of 303k. In Figure 5, the solubility curves of mandelic acid under high pressure obtained in this study are shown. The solubility values of racemic compound which was obtained in the previous study (2) are also plotted in the figure for other temperatures. The solubility curve of mandelic acid at an atmospheric pressure (0.098 MPa) at 303K is also shown in the same figure. The plots of the solubility obtained from an analysis of the mother liquor removed from the feed A, B, and C form smooth curves.

The L-form composition in eutectic solution is constant at 70% at atmospheric pressure as seen in Figure 5. It shifts to the racemic side when the pressure is elevated (compare Figure 4 and Figure 5). It is clear that the region of chiral formation becomes wider at elevated pressure. In these experimental conditions, the zone of racemic compound formation didn't disappear but only diminished. The mandelic acid in mother liquor separated from the feed C (L-70%) was 62 to 64% in its L-form concentration and the concentration were higher than the solubility of both the chiral and the racemic of mandelic acid. The analytical results showed that the obtained crystals contained both L-form and D-form. Furthermore, XRD analysis of these crystals revealed that L-form and D-form in these crystals are derived from L-form crystals and from the racemic crystals respectively.

These results indicate that the mother liquor of saturated mandelic acid solution may coexist with both L-form crystals and the racemic crystals, when feed C was pressurized in the vessel for crystallization. The coexistence of the chiral and the racemic indicates that the composition of the solution was the eutectic. So, the solubility and the L-form composition of mandelic acid in the mother liquor represents the eutectic points of mandelic acid under pressure (see the arrow from point C in Fig.5).

Results of L-form separation

L-form separation from feed in Zone Ⅰ   The mother liquor obtained from feed A (L-90%) and B (L-85%) in the experiment of solubility measurement is considered to be the saturated solution corresponding to each operational condition. The amount of D- and L- mandelic acid in the mother liquor obtained from the experiments of feed A and B are plotted against pressure in Figure 6 and Figure 7, respectively. In these figures, the amount of D-form remained in the mother liquor was almost constant. From this, it is hypothesized that only L-form crystal could have been crystallized. The composition of L-form and the yield of the crystal obtained from feed A and B are plotted against pressure in Figure 8 and Figure 9. Highly pure crystals of L-mandelic acid, 99.3% and 98.5% purity, were obtained from feeds A and B, respectively. The yield became somewhat higher with the increase in pressure. The compositions of L-form in the obtained crystals were higher than the calculated values from the solubility curves. This is because the crystals were purified by sweating and recrystallization mechanisms accompanied with a little pulsation of liquid pressure during transferring the liquid. These facts show a possibility of obtaining crystals with high purity from the feed in Zone Ⅰ (Step 2).

L-form recovery from feed in Zone Ⅱ   The composition of L-form in crystals, the crystal yield and the L-form composition of the mother liquor from feed C (L-70%, eutectic composition) are shown in Table Ⅰ. L-mandelic acid of 79.2% max. purity was obtained, so these crystals can be fed to the separation step (Step 2) in Zone Ⅰ for further purification. The weights of the

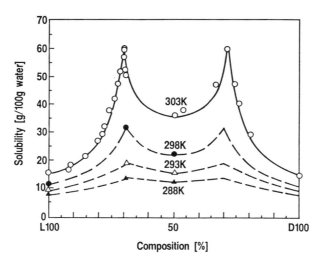

Figure 4. Solubility diagram of D-, L-mandelic acid at atmospheric pressure (0.098 Mpa)

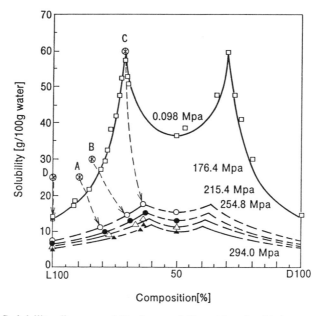

Figure 5. Solubility diagram of D-, L-mandelic acid under high pressure (303K)

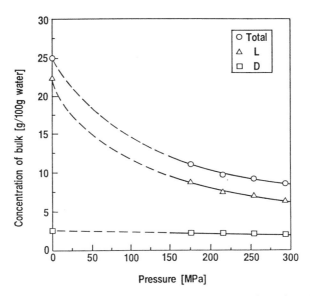

Figure 6. Relationship between pressure and concentration of mother liquor (Feed A : 25g/100g water, L90%, 303K)

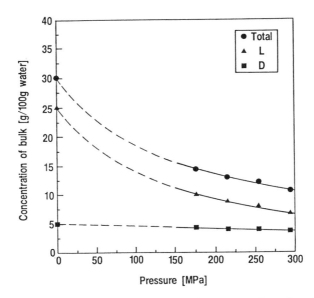

Figure 7. Relationship between pressure and concentration of mother liquor (Feed B : 30g/100g water, L85%, 303K)

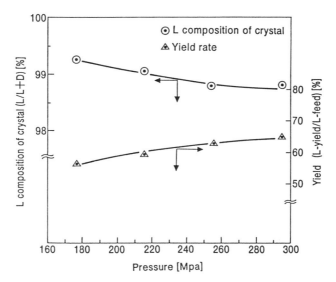

Figure 8. L-concentration of the obtained crystal and yield as a function of pressure (Feed A : 25 g/100g water, L/D=90/10, 303K)

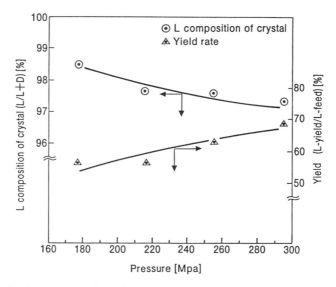

Figure 9. L-concentration of the obtained crystal and yield as a function of pressure (Feed B : 30 g/100g water, L/D=85/15, 303K)

obtained crystals were somewhat lower than those calculated from the solubility curves, especially in the lower pressure region. This is because the entire amount of crystals couldn't be recovered and portion of the crystal was lost due to melting caused by pulsation of liquid pressure in the vessel during liquid-solid separation. Also the calculated value became larger than the real value, because of the decrease in concentration of mother liquor due to water contamination in the pipe or the vessel entering the feed solution. However, in contrast, the purity of L-form was higher than the calculated value in the lower pressure region, because of the melting mentioned above.

Discussion

A proposed separation process of L-form crystal from L-rich D-,L-mandelic acid mixture is presented. This is based on the experimental results described above and the idea shown in Figure 2. This process consists of three steps. In Step 1, the feed mixture is separated by high pressure crystallization (1) from the eutectic mixture at atmospheric pressure. In Step 2, L-form crystals with high purity are obtained from this feed using high pressure crystallization (2). In Step 3, the eutectic mixtures removed by high pressure crystallization (1 and 2) is concentrated to a nearly eutectic composition at atmospheric pressure, by means of evaporation and cooling crystallization.

For example, the eutectic mixture (Zone II) in the L-rich region under atmospheric pressure is assumed to be a feed sample. At first, the feed is crystallized by high pressure crystallization (1) to separate out L-form crystals with a purity of more than 70% (Zone I). In the next step, an aqueous solution of mandelic acid is made up by adding water to the crystals thus obtained. The composition of this solution is adjusted so that L-form composition is larger than the value of S × (L/L+D) and D-form composition is also smaller than the value of S × (D/L+D). Here, S indicates the eutectic solubility at the corresponding operational condition. By applying high pressure crystallization (2) to this feed solution, only L-form crystal can be separated out from the L-rich asymmetric solution of D-, L-mandelic acid mixture. The composition of L-form in the mother liquor (Zone III) separated by this high pressure crystallization(2) comes out to be nearer to the racemic than the eutectic at atmospheric pressure.

The concentration step by evaporation and cooling crystallization has not been studied experimentally, as it should be accomplished easily by conventional methods. The solution near the eutectic composition (L/D=70/30) at atmospheric pressure is made up by concentrating the separated mother liquor in the previous step (Step 1 or Step 2) and crystallizing the racemic crystal. This solution is used again as feed for Step 1.

Conclusions

The following facts were clarified, in relation to the separation of L-form crystals from L-rich D-,L-mandelic aqueous solution (see also Table II).
1. The eutectic point in the D-,L-mandelic aqueous solution system moves toward the racemic composition under high pressure.
2. High purity L-form crystals of more than 99% can be separated from the feed which contains more L-form than the eutectic composition under high pressure, and the purity is a little higher than the value calculated from solubility curves.
3. The possibility of a process which can separate L-form crystals from L-rich D-,L-mandelic aqueous mixture, by combining high pressure crystallization and cooling crystallization has been clarified.

Table I. Composition of L-form in crystal and in mother liquor (Feed C : 60g/100g water, L/D=70/30, 303k)

	Pressure MPa	Crystal yield		Composition of L-form in crystal		Composition of L-form in mother Liquor	
		$\frac{(D+L) crystal}{(D+L) feed}$ %	$\frac{L crystal}{L feed}$ %	Measured %	Calculated %	Measured %	Obtained from solubility %
1	176.4	34.2	38.8	79.19	72.6	63.53	63.5
2	215.6	41.7	45.0	75.48	72.6	63.14	62.3
3	254.8	48.1	50.7	73.81	72.6	63.26	61.3
4	294.0	66.2	68.0	71.91	72.6	62.81	60.0

Table II. Summary of results

Zone	Feed	Means	Results	
			Crystal	Mother Liquor
I	≫ 70%	High Pressure Crystallization	L-form (99%up)	Zone II or Zone III
II	⋮ 70%	High Pressure Crystallization	Zone I (72~80%)	Zone II (62%)
III*	＜ 70%	Evapolation & Cooling Crystallization	Racemic Compound	Zone II

* not studied experimentally

Literature Cited

(1) Moritoki, M.; Kitagawa, K.; Onoe, K.; Kaneko, K. Industrial Crystallization '84; Jancic, S.J.; de Jong, S.J., Ed.; Elsevier Sci. Publishers B. V: Amsterdam, 1984, pp 377-38.
(2) Shinohara, T. Master's Thesis (Waseda University), 1995.

Chapter 8

Morphological Change Mechanism of α-L-Glutamic Acid with Inclusion of L-Phenylalanine

M. Kitamura

**Department of Chemical Engineering, Hiroshima University,
1–4–1 Kagamiyama, Higashi-Hiroshima 739, Japan**

The effect of L-Phenylalanine(L-Phen) on the growth rate and morphology of α crystal of L-glutamic acid(L-Glu) was investigated using a single crystal method in a flow system. The growth rates in a directions parallel to the main plane of the crystal, A and B (corner) were suppressed by L-Phen and the rates decreased with increasing concentration of L-Phen. In addition the growth rate got to fluctuate with time. On the other hand, the growth rate in the direction of thickness(D), corresponding to (001) face, was scarcely affected by L-Phen even at much higher L-Phen concentration. In the growing process a new face was found to appear, and the face index was determined as (110). This means that L-Phen adsorbs preferentially on (110) face in comparison with (111) and (001) faces. The fluctuation of the growth of (110) was attributed to the inhomogeneity of the crystal surface. Furthermore, it was shown that L-Phen adsorbed on a crystal surface is included in α L-Glu crystals in the growing process.

Problems caused by polymorphism appear in many fields such as fine chemicals in industries (pharmaceuticals(*1,2*), foods, etc.), optical electronic materials(*3*), clathrate compounds(*4*) and biominerals(*5*). In crystallizations of these materials the crystallization behavior of the polymorphs is controlled first by basic operational conditions such as temperature, supersaturation degree, stirring rates. In addition to these basic factors, solvents, additives and guest molecules (in clathrate compounds) should be also considered as the important factors(*6,7*). The crystallization process of the polymorphs is composed of competitive nucleation, competitive growth of polymorphs and transformation from metastable to stable form. Accordingly individual step should be investigated to clarify the crystallization mechanism of polymorphs.

On this account we have been investigating the polymorphism of L-glutamic acid(*7-9*). L-Glutamic acid has two polymorphs a metastable and a stable β, which have different lattice parameters of the same space group of Orthorhombic P2$_1$2$_1$2$_1$(*10,11*). We reported previously that at 45°C in pure solutions both polymorphs crystallize and the transformation by a solution-mediated mechanism occurs (8). When L-Phen was present the crystallization behavior of the polymorphs was influenced remarkably(*12*). On the other hand, at 25°C in pure solution, only α crystals tended to crystallize out and the transformation could not be observed(*8*). However, it appeared that L-Phen retards the nucleation and crystallization rates of α crystals, and induces the morphological change of α crystals.

In this study, the effect of L-Phen on the growth kinetics and morphological change of α L-Glu was investigated at 25 °C with a single crystal method in a flow system.

Furthermore, inclusion of L-Phen in α L-Glu crystals was also examined by a batch crystallization.

Experimental apparatus and procedure

In Figure 1 the apparatus for a single crystal method is shown. L-Glu and L-Phen were dissolved in distilled water in the tank(1) and the solution was circulated with a pump through the heat exchanger(3) to the growth cell(4). The size of the crystals were measured by the microscope and video-TV system. The velocity of solution in the cell was varied between 0.05 and 0.25 m/s.

The seed crystals of α and ß prepared by batch crystallizations in solutions without L-Phen were set in the growth cell as shown in Figure 2. The flow direction is shown with an arrow. The crystal size changes (ΔL) in the directions of A1, A2, A3 and B, which are parallel to the main face and perpendicular to each edge face were measured. The increment, ΔL in the directions of D1 and D2 (thickness) was also measured.

Inclusion of L-Phen in L-Glu was also examined by batch crystallization at 25°C. Crystallization of L-Glu was done by rapid cooling method at various L-Phen concentration, keeping the L-Glu concentration constant at 0.204 mol/l. The amount of L-Phen in L-Glu crystals was determined by liquid chromatography.

Results and discussion

Growth of L-Glu in pure solutions

In Figure 3 the typical results of the relationship between the size change , ΔL, and the elapsed time t in A and B directions in pure solution are shown (L-Glu concentration is 0.091 mol/l; solution velocity, v, is 0.148 m/s). Linear relationship could be obtained and from the slope the growth rates were determined. It was observed that the slopes in the A1,A2 and A3 directions, G(A), were very similar and seem to be larger than that in B direction. These results support a tendency for developing a face in B direction during the measurement.

In Figure 4 the dependence of the size change , ΔL on time t in D direction is shown. The L-Glu concentration is the same as that in Figure 3. Good linear relationship appears again and the growth rate in D direction was estimated as about 60 % of that in A direction. The ratio of the growth rates in A and D directions coincide with the plate-like morphology of the α crystals.

Effect of solution velocity on the growth rate

It can be considered that the growth process is composed of the following steps in series: the volume diffusion of the solute through the boundary layer and the surface reaction process including adsorption of the solute on the crystal surface, surface diffusion of the solute and integration at the kink. The mass transfer coefficient is usually proportional to $v^{1/2}$ (v is solution velocity) if flow in boundary layer is laminar. In Figure 5 the dependence of growth rate in A direction on solution velocity, v is shown. In the velocity range of our experimental study, little effect was observed on the growth rates in A and D directions. The results indicate that the growth process in both directions is mainly controlled by the surface reaction process and the contribution of the diffusion process in solution is small.

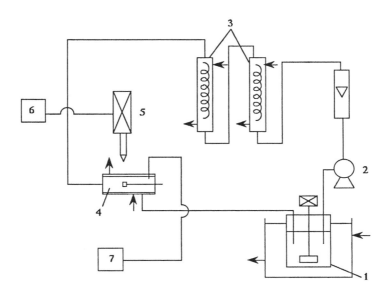

1. Dissolving tank   5. Microscope
2. Pump   6. TV Camera and Video
3. Heat exchanger   7. Thermocouple
4. Growth cell

Figure 1 Experimental apparatus of a single crystal method.

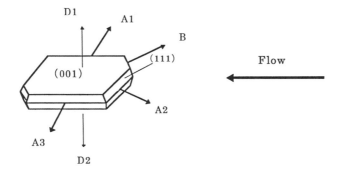

Figure 2 Seed crystal of α L-Glu

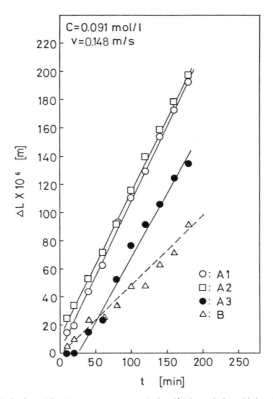

Figure 3 Relationships between crystal size(ΔL) and time(t) in A direction

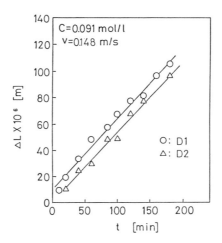

Figure 4 Relationships between crystal size(ΔL) and time(t) in D direction

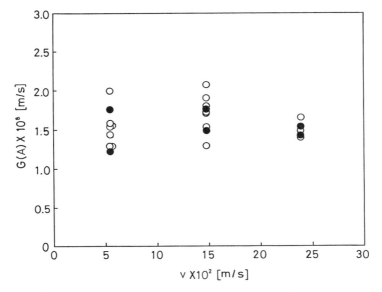

Figure 5 Relationship between the growth rate and solution velocity in A direction; black circle indicates especially the results in A1 direction

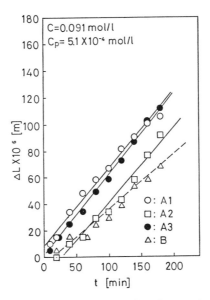

Figure 6 Dependence of crystal change in A direction on elapsed time at L-Phen concentration of 5.1×10^{-4} mol/l.

Effect of L-Phen on Growth of L-Glu

When L-Phen was added in the solution, the growth rates in A and B directions decreased with increasing L-Phen concentrations.   In Figure 6 the results for the case of L-Phen concentration of 5.1×10^{-4} mol/l is shown.   The growth rates are about 70 % of that in pure system.   It appeared that as the suppression of the growth rate with L-Phen is larger for A than B direction, the face of B tended to disappear.

In this concentration range the linear relationship can be seen, however, at higher L-Phen concentrations, the linear relationship between ΔL and Time was not established.   The growth became much irregular and the growth rate fluctuated between intervals of no growth and intervals of growth.

From these results the growth rates were calculated and in Figure 7 the dependence of the growth rate of G(A) on L-Phen concentration (Cp) is shown.   The dispersion of the values of G(A) depending mainly on seed crystals and time are shown by vertical bars.   However, it can be seen that the growth rate decreased clearly with increasing L-Phen concentration and stopped completely at 1.8×10^{-3} mol/l (Cp).   Such decrease of the growth rate by an additive was observed by us also for ammonium sulfate in the presence of chromium (III) ion(13).

On the other hand, in D direction no effect was be observed.   In Figure 8 the result at L-Phen concentration of 2.6×10^{-3} mol/l is shown.   Even when the concentration of L-Phen(Cp) was increased up to 5.0×10^{-3} mol/l, the decrease of the growth rate could not be detected.   These results indicate that the shape of L-Glu crystal becomes thicker with increasing L-Phen concentration.

Morphological change

The results mentioned above mean that in the presence of L-Phen the crystals grow preferentially in D direction ([001]), preferentially grow .   It was expected that the growth of (111) face is inhibited by L-Phen.   However, microscopic observations indicate that a new face appears as shown in Figure 9.   With X-ray diffraction analysis, the new face was decided to be (110).   These results suggest that L-Phen may be adsorbed on (110) face selectively and inhibit the growth of that face in comparison to (001) and (111) faces.

Mechanism of L-Phen effect on morphological change and growth process of L-Glu.

Growth rate in A direction can be attributed to that of the (110) face.   When the structure of the (110) surface was compared with that of (001) face from crystallographic data, it became clear that the (001) face is very rich in carboxylic acid group.   This suggests that phenyl group of L-Phen may be repelled by the (001) face and on this account L-Phen molecule can not be adsorbed on the (001) face.   On the other hand the molecular arrangement of L-Glu in the (110) face seems to be partly suitable for the adsorption of L-Phen, i.e. L-Phen can be adsorbed on the surface of the (110) through the common part of amino acid including the chiral carbon(14).

It was indicated that L-Phen adsorb preferentially on the (110) face and the growth rate of that face was not only suppressed but also it fluctuated with time as described previously.   Such fluctuation suggests that the adsorption of L-Phen on the crystal surface is inhomogeneous.   We observed previously a similar decrease of the growth rate for ammonium sulfate in the presence of chromium (III) ion(13).   However, in that case the growth rate progressively decreased with time and did not fluctuate. Such difference may de related to the difference of the growth mechanism and characteristics of the surface of each material.

Inclusion of L-Phen in α L-Glu crystal

Inclusion of L-Phen in L-Glu crystals is an interesting problem because it relates with the mechanism of the additive effect.   It was examined by batch crystallization

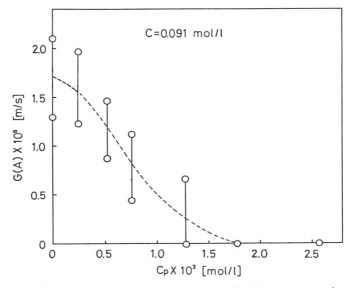

Figure 7 Relationship between the growth rate and L-Phen concentration in A direction

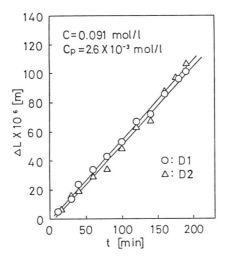

Figure 8 Dependence of crystal change in D direction on elapsed time at L-Phen concentration of 2.6x10^{-3} mol/l in the presence of L-Phen

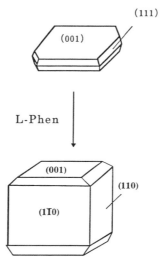

Figure 9 Morphological change of L-Glu seed crystal.

at 25°C at higher concentration ranges of L-Glu and L-Phen than those in the single crystal method. The concentration of L-Glu was kept constant as 0.204 mol/l and the L-Phen concentration was changed between 10 - 40x10^{-3} mol/l. In Table I the amounts are shown as the molar ratio to L-Glu crystals(R). It is clear that L-Phen is included in L-Glu crystals and the amount increases with increasing L-Phen concentration in solutions. This indicates that the L-Phen adsorbed on the crystal surface is included in the crystal in the growing process of L-Glu.

Table I. Inclusion of L-phen in L-Glu crystals (L-Glu conc. = 0.204 mol/l)

L-Phen conc., x10^3 mol/l	Molar ratio, L-Phen/L-Glu in crystals [-]
10.3	0.0042
20.5	0.0074
30.8	0.0133

Conclusion

The growth rates of α L-Glu in A and B directions decreased   remarkably with increasing concentrations of L-Phen and the growth got irregular at higher L-Phen concentrations. On the other hand, the growth rate in the D direction, corresponding to (001) face, was scarcely affected by L-Phen even at much higher L-Phen concentration. In the growing process a new face was found to appear, and the face index was determined as (110). This means that L-Phen is adsorbed preferentially on (110) face in comparison to (111) and (001) faces. The fluctuation of the growth of (110) was attributed to the inhomogeneity of the crystal surface. It was found that a slight amount of L-Phen is included in the L-Glu crystals and the amount increased with increasing L-Phen concentration. This means that the L-Phen adsorbed on the crystal surface is included in growing crystals of L-Glu.

Literature cited

1. Borka, L., *Acta Pharm. Suec.*, **1974**, 11, 295
2. Imaizumi, H. ; Nambu, N.; Nagai, T, *Chem. Pharm. Bull.*, **1980**, 28, 2565
3. Hall S.R., et al, *J. Crystal Growth*, **1986**, 79, 745
4. Kitamura, M., *J.Crystal Growth*, **1990**, 102, 255
5. Garside J.; Brecevic, Li; Mullin, ,J.W., *J. Crystal growth*, **1982**, 57, 233
6. Kitamura, M., *J. Soc. Powder Technology Japan*, **1992**, 29, 118
7. Kitamura , M., *J. Jap. Assoc. Crystal Growth*, **1989**, 16, 61
8. Kitamura , M., *J.Crystal Growth*, **1989**, 96, 541
9. Hiramatsu S., *Nippon Nogeikagakukaishi*, **1977**, 51, 27
10. Hirayama N.,K.Shirahata, Y.Ohasgi and Y.Sasada, *Bull. Chem. Soc. Jpn.*, **1980**, 53, 30
11. Hirokawa , S., *Acta Cryst.*,   **1955,** 8,637
12. Kitamura, M. ; Funahara H., *J. Chem. Eng. Japan*, **1994**, 27, 124
13. Kitamura, M.; Kawamura, Y.; Nakai, T., *Inter. Chem. Eng.*, **1992**, 32, 157
14. Staab E., L.Addadi, L.Leiserowitz and M.Lahav, *Adv. Mater.*, **1990**, 2, 40

Chapter 9

Growth Process and Morphological Change of β-Glutamic Acid in the Presence of L-Phenylalanine

M. Kitamura and Y. Sumi

Department of Chemical Engineering, Hiroshima University, 1–4–1 Kagamiyama, Higashi-Hiroshima 739, Japan

The growth rate of ß crystal of L-glutamic acid(L-Glu) was measured using a single crystal method in a flow system at 25°C. The growth rate in the longitudinal direction(A), which corresponds to (101) face, was measured under the experimental conditions, however, the growth rates in the directions of thickness(C) and width(B) were too low to be observed. The growth process in pure solutions was found to be controlled by the surface reaction step. It was observed that with increasing of L-phenylalanine(L-Phen) concentration (Cp) the growth rate decreased and finally stopped at the critical concentrations (Cp*). At higher concentrations of L-Phen, the growth of L-Glu became irregular and a fluctuation of the growth rates with time appeared. The fluctuation was attributed to the complex surface structure, i.e. on the surface adsorption of L-Phen is not homogeneous and the growth of L-Glu is occurring competitively. Furthermore, a morphological change occurred, i.e. the growing face of (101) tended to separate. It is considered that the adsorption of L-Phen occurs at a specific point on the advancing front of the crystals. When the value of Cp* for ß crystal was compared with that for α crystal in the direction parallel to the main face. it was cleared that Cp* for ß is smaller than that for ∝ . This fact coincides well with the batch crystallization behaviors of these polymorphs.

In the crystallization of polymorphs influence with some special additives is observed(1-3). Such effect is due to the relative change of nucleation rate and growth rate of the polymorphs,' the effect also contains the change in transformation rate(4,5). Furthermore, a morphological change occurs in many cases. On the other hand, a crystallographic approach has been applied to effect of various type of additives on crystal growth(6-8). However, the effect of additives especially on the kinetic behavior of polymorphism has not usually been treated quantitatively and the mechanisms are not known.

   From this pointof view we have dealt with the crystallization and the transformation processes of amino acids such as L-glutamic acid(L-Glu)(9.10) and L-histidine(11). With respect to L-Glu, the effects of temperature, supersaturation, and stirring on the crystallization behavior of the polymorphs(metastable α and stable ß) were clarified(9). At 45°C in pure solutions both polymorphs tended to crystallize and the transformation occurred by "solution-mediated mechanism".   However, in the presence of L-phenylalanine (L-Phen) the crystallization of ß type was selectively suppressed and the transformation rate was retarded(*10*).   This means that L-Phen suppresses preferentially both the nucleation rate and growth rate of ß L-Glu.

In this work the effect of L-Phen on the growth kinetics and morphology of ß L-Glu crystals was investigated using a single crystal method in a flow system and the results were compared with those for ∝ L-Glu crystals.

Experimental apparatus and procedure

The same apparatus, which is described elsewhere(*12*), is used for the measurement of crystal size change. The size of the crystals were measured by the microscope and video-TV system. The velocity of solution in the cell was varied between 0.05 and 0.25 m/s.

The seed crystal prepared by batch crystallization in solutions without L-Phen was set in the growth cell as shown in Figure 1.   The flow direction is shown by an arrow.

The shape of ß crystals is that of a long plate.   The crystal size change(ΔL) in A (length), B(width) and C(thickness) directions were measured.   In A direction the displacement of a fixed point on an edge was measured.   The growth rate in A direction, G(A) can be reduced to the growth rate of (101) face, as the angle θ in Figure 1 is known (θ =35°). The growth rate measurements of these polymorphs were carried out first in pure solutions(without L-Phen)   and secondly in solutions containing L-Phen at various concentrations.

Results and discussion

Growth in pure solutions

In Figure 2 typical results of the dependence of the size change , ΔL on elapsed time t in pure solution are shown (L-Glu concentration is 0.078mol/l).   No change of size was observed in B and C directions.   Only in A direction a change could be detected. Linear relationship was obtained between the size change and time, and from the slope the growth rate was estimated.   When the concentration was increased to 0.093 mol/l, the linear relationship also held for the A direction and the slope of the line increased, corresponding to a higher growth rate(Figure 3).   However, in B and C directions no size change could be observed.

To examine the contribution of the diffusion process of the solute in the growth process of ß L-Glu the dependence of the growth rate in A direction on solution velocity(v)   was measured. In Figure   4   the results obtained at different supersaturation degrees are shown.   It can be seen that in both supersaturation degrees, almost no influence of the solution velocity are observed.   The results indicates that the growth process is controlled by the surface reaction process and the contribution of diffusion process is small.   The growth mechanism for α crystal was also confirmed to be mainly controlled with surface reaction process(*12*).

Growth in solutions containing L-Phen

When L-Phen was added in the solution the growth of the crystal in A direction was suppressed.   In Figure 5 the dependence of crystal size change on time at L-Phen concentration of 3.85 x10^{-4} mol/l is shown (L-Glu concentration is 0.093 mol/l). The marks in the figure indicate the results for two different crystals.   The growth rate estimated from the slope appeared to be about 70 % of that in pure solution in Figure 3.

When the concentration of L-Phen was further increased no linear relationship between ΔL and time was seen any more and the growth became very irregular. In

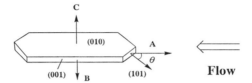

Figure 1 Seed crystal set in growth cell.

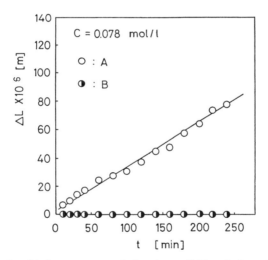

Figure 2 Relationship between crystal size change(ΔL) and elapsed time (t) at L-Glu concentration of 0.078 mol/l.

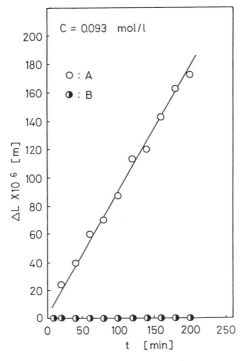

Figure 3 Relationship between crystal size change(ΔL) and elapsed time (t) at L-Glu concentration of 0.093 mol/l.

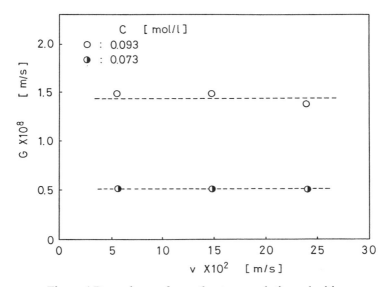

Figure 4 Dependence of growth rates on solution velocities.

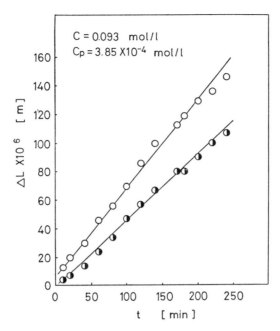

Figure 5 Dependence of crystal size change on time at L-Phen concentration(Cp) of 3.85 x10-4 mol/l (Marks correspond to two different crystals) .

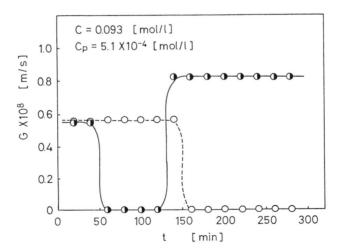

Figure 6 Fluctuation of the growth rate(G) with elapsed time (Marks indicate two different crystals).

Figure 6 the typical results of the growth behavior are shown(L-Phen concentration is 5.1×10^{-4} mol/l). The marks indicate the growth rates of two different crystals. It is clear that the growth of the crystals fluctuate with time, i.e. the crystal abruptly stopped growing and after some interval it began growing again. Such irregular growth behavior depends on the crystal and also positions of the growing front of the crystal.

The growth rate was calculated for each run. The values of the growth rates were dispersed with crystal and time as mentioned in Figure 6.   In Figure 7 the dependence of the growth rate in A direction($G(A)$) on L-Phen concentration(Cp) is shown and the dispersion of the growth rate is denoted with bars. However, t can be seen that with increasing L-Phen concentration (C_p) the growth rate decreased and finally stopped at a critical concentration (C_p^*).

Furthermore, in the course of the growth of the crystal a morphological change was observed, i.e. the growing face of (101) tended to separate(Figure 8). As a reason for the separation of the growing front, we considered the possibility that at specific point on the advancing front, an intense adsorption of L-Phen occurred and the growth of that point was inhibited.

Surface structure of (101) face
It is clear from the results that L-Phen adsorbs on (101) face and inhibits the growth of ß crystals. The molecular arrangement on (101) face was examined closely as shown in Figure 9. It can be seen in Figure 9 that the L-Glu molecules seems to be in a line along the (101) surface. Accordingly L-Phen should attack from the side of the line of L-Glu. However, in the L-Glu arrangement two kinds can be distinguished. The one molecular arrangement is considered to make possible the adsorption of L-Phen through the common part of amino acid(8).

Comparison of the effect of L-Phen between α and ß crystals.
It was observed that L-Phen retards also the growth rate of α polymorph of L-Glu in the direction(A) parallel to the main face of α crystal(10). However, no effect could be detected in the direction of the thickness of α crystal. On the other hand, as mentioned above, the growth of ß crystals in A direction, which corresponds to the main growing face(101), is remarkably inhibited by L-Phen. When the critical concentrations for L-Phen (C_p^*), at which the growth is completely stopped, are compared between α and ß crystals, it is clear that the value of C_p^* for ß in A direction is about half of that for α in A . This fact indicates that the effect of L-Phen on the (101) faces of ß crystal is stronger than that on the (110) faces of α crystal. Furthermore, these results coincide with the crystallization behavior of the polymorphs in batch crystallization at 45°C, i.e. α crystals precipitate preferentially in the presence of L-Phen(10).

Mechanism of fluctuation of growth rates
As shown previously, L-Phen as an additive makes the growth rates fluctuate remarkably. This seems mean that the growing face is not smooth at microscopic scale. During the growth of the crystal the real surface of crystals may be complicated because advancing of the steps and the adsorption of L-Phen occur competitively(Figure 10). As a result the density of L-Phen adsorbed on the crystal surface is not uniform all over the crystal surface. According to the model of Cabrera and Vermylea(11) the advancing step has to jut out like a bow between the additive molecules adsorbed on the crystal surface. It is considered that the growth rate is not same all over the surface and varies dependently on the density of L-Phen on the

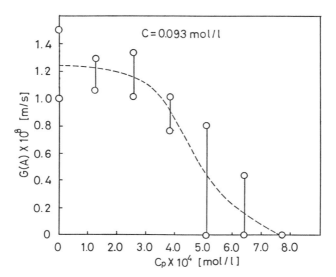

Figure 7 Dependence of the growth rate on the concentration of L-Phen(Cp).

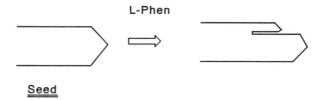

Figure 8 Morphological change in growing process of ß crystal.

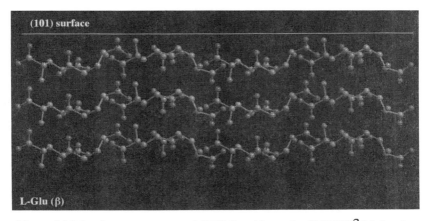

Figure 9 Molecular arrangement of (101) face (drawn by SERIUS[2](Molecular Simulations Inc.))

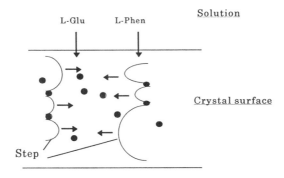

Figure 10 Irregular advance of the step in the presence of L-Phen.

crystal surface. We have observed previously a decrease of the growth rate for ammonium sulfate in the presence of chromium (III) ion(*14*). However, in that case the growth rate continuously decreased with time and the fluctuation was not detected. Such differences may be related to differences in the type of the adsorption of additive and the growth mechanism of each crystal.

Conclusion

The growth rate in the longitudinal direction(A), which corresponds to (101) face, was measured under the experimental condition., however, in the directions of thickness(C) and width(B) the measurement was not possible because it was too low to be observed. The growth process in pure solutions was confirmed to be mainly controlled by a surface reaction step. With increasing of L-Phen concentration , the growth rate decreased and finally stopped at the critical concentrations (C_p^*). At higher concentrations of L-Phen, the growth process got irregular and a fluctuation of the growth rates with time appeared. The fluctuation was attributed to the complicated surface process, i.e. on the surface adsorption of L-Phen is not homogeneous and the growth of L-Glu is occurring competitively. In the growth process a morphological change occurred, i.e. the growing face of (101) tended to separate. It is considered that a characteristic adsorption occurs at a specific point on the advancing front of crystals. The critical concentration(C_p^*) for ß is about half of that for α crystal in the direction parallel to the main face. This fact coincides well with the batch crystallization behaviors of these polymorphs. The result indicates that the effect of L-Phen on (101) face of ß crystal is stronger than that on (110) face of α crystal.

Literature Cited

1.Z.Kolar ; Binsma, J.J.M; Subotic, B, *J.Crystal Growth*, **1984**, 66, 179
2. Hall S.R., et al, *J. Crystal Growth*, **1986,** 79, 745
3.Kuroda K.; Yokota,T.; Umeda,T; Kita,Y; Konishi, A; Kuroda,t; *Yakugaku Zasshi*, **1979,** 99, 745
4. Kitamura, M., *J. Soc. Powder Technology Japan*, **1992**, 29, 118
5. Kitamura , M., *J. Jap. Assoc. Crystal Growth*, **1989**, 16, 61
6. Staab E., Addadi, l.; Leiserowitz, l; Lahav, m., *,Adv. Mater.*, **1990**, 2, 40
7. Van Rosmalen, G.M; Bennema, P, *J. Crystal Growth*, **1990**, 99, 1053
8 Addadi,L., Berkovitch-Yellin, Z., Weissbuch,I., Mil van J., Simon, L., Lahav,M. and Leiserowitz, l., *Angew. Chem. Int. Ed.Engl.,* **1985**, 24, 466
9. Kitamura , M., *J.Crystal Growth*, **1989**, 96, 541
10. Kitamura, M. ; Funahara H., *J. Chem. Eng. Japan*, **1994**, 27, 124
11. Kitamura, M.; Furukawa, H.; Asaeda S., *J.Crystal Growth*, **1994,** 141, 193
12. Kitamura M., submitted to ACS symposium series.
13.Cabrera , N.; Vermilyea, D.A., *Growth and Perfection of Crystals*; Weily & Sons, 1958; pp.393-407
14. Kitamura, M.; Kawamura, Y.; Nakai, T., *Inter. Chem. Eng.*, **1992**, 32, 157

Chapter 10

Crystallization Behaviors of α- and β-Quizalofop-ethyl Polymorphs in Homogeneous Nucleation

K. Miyake, O. Araki, and M. Matsumura

**Department of Chemical Engineering, Hiroshima University,
1–4–1 Kagamiyama, Higashi-Hiroshima 739, Japan**

The induction time in homogeneous nucleation of α or β polymorphs of quizalofop-ethyl, i.e. the time the crystals first appeared in a supersaturated solution of quizalofop-ethyl-ethanol mixtures was measured at temperatures between 293 and 311K in a batch stirred crystallizer. Experimental equations on the induction time have been derived semiquantitatively on the basis of an equation of classical nucleation rate. In addition, the concentration of supersaturated solution $C_{\alpha\beta}$, when both forms of α and β simultaneously occurred, has been calculated from the experimental equations of the induction time, and is given by the equation:

$$C_{\alpha\beta} = 6.2 \times 10^{-15} \exp(0.1005\,T)$$

where, T is Kelvin temperature. The equation can be applied at temperatures between 305 and 311K.

Quizalofop-ethyl($C_{19}H_{17}O_4N_2Cl$: hereafter abbreviated as QE) a useful material for the production of herbicides it is an optically active compound and has two known polymorphic forms:α (metastable type) and β (stable type) *(1- 4)*. The shape of α is needle or prism-like, and that of β is filamentary crystals *(2), (4)*. The effect of QE as a herbicide is the same regardless of the polymorph. However, the filamentary crystals of β has a negative effect on the stirring. The crystals aggregate and entwine the impeller of the agitator in the crystallizer. In addition, the filamentary crystals present problems in filtration and drying because of the solvent they are with holding.

Consequently, the selective production of crystals of α metastable type is desirable. Accordingly, it is necessary to study the crystallization behavior of the polymorph appearing first (primary nuclei) in a supersaturated solution.

In this paper, correlating equations of solubility of QE in ethanol were proposed using previous data *(2)*. Subsequently, in order to clarify the formation mechanism of polymorphs appearing first in a supersaturated solution, the solutions were cooled to a given temperature rapidly and the time required for crystallization (induction time) was measured. The discrimination between the α and β forms in the appearing crystal was also done. The induction time of each polymorph was interpreted semiquantitatively on the basis of classical equation of nucleation rate *(5 - 6)*.

Correlating equations of solubility

The author et al. *(2)* measured the solubility of α and β polymorphs of QE in ethanol at temperatures between 273 and 317K. From these data *(2)*, the following correlating equations for the solubility were obtained.

For α

$273 \leqq T \leqq 284$

$$C_{s\alpha} = 7.811 \times 10^{-4} T - 0.20412 \qquad (1)$$

$284 \leqq T \leqq 305$

$$C_{s\alpha} = 9.75 \times 10^{-13} \exp(8.123 \times 10^{-2} T) + 0.00753 \qquad (2)$$

$305 < T < 317$

$$C_{s\alpha} = 3.41 \times 10^{-13} \exp(8.5 \times 10^{-2} T) \qquad (3)$$

For β

$273 \leqq T \leqq 284$

$$C_{s\beta} = 6.84 \times 10^{-4} T - 0.1765 \qquad (4)$$

$284 \leqq T \leqq 305$

$$C_{s\beta} = 7.93 \times 10^{-13} \exp(8.14 \times 10^{-2} T) + 9.03 \times 10^{-3} \qquad (5)$$

$305 < T < 317$

$$C_{s\beta} = 2.85 \times 10^{-20} \exp(0.1353 T) + 3.425 \times 10^{-2} \qquad (6)$$

where $C_{s\alpha}$ [kg-solute /kg-EtOH] is the concentration of saturated solution for α at a given temperature. $C_{s\beta}$ [kg-solute/kg-EtOH] is the concentration of saturated solution for β at a given temperature.

The empirical equations of these solubilities agree with 99% of the previous data *(2)* within ± 6%.

Experimental apparatus

A schematic diagram of the experimental apparatus is shown in Figure 1. This apparatus has some similarity to the apparatus used in the previous paper *(7)*. A glass stirred tank crystallizer ① with 125 ml effective volume was used, which had a flat plate impeller wing of Teflon (wing diameter 39mm, width 11mm) located near the bottom and a Pt resistance thermometer of 1.6mm OD ③ for measuring solution temperature. The rotational speed of wing ② was 400 ± 5 rpm throughout the experiments. A cold light of halogen lamp irradiated from behind the crystallizer. The crystals first appearing in the solution were observed with the help of this light. The solution temperature in the crystallizer was adjusted by three water baths with circulation pumps ⑤,⑥ and⑦. These water baths used mixtures of ethylene glycol-water as the heating medium.

Procedure. The α QE crystals of purity 99.5% were supplied by Nissan Chemical Industries, Ltd., and were used without further purification. The ethanol used as solvent was of special grade reagent. These chemicals were weighed by an electronic precision balance in order to prepare a fixed concentration solution. The volume of the fixed concentration solution fed into the crystallizer was 125 ml. The temperature of the heat medium in the first water bath ⑤ was kept at a temperature about 5K higher than the saturated temperature of the preparated solution. After the heat medium was circulated for 20 minutes to the jacket of the crystallizer, the solute in the crystallizer was completely dissolved. On the other hand, the heat medium in the second water bath ⑥was kept at a temperature about 15K lower than the crystallization temperature. After the heat medium of high temperature had circulated for 20 minutes to the crystallizer, the cold heat medium of the second water bath was circulated to the jacket of the crystallizer, and the solution in the crystallizer was cooled quickly from the dissolving temperature of solute to crystallization temperature. Therefore, the solution in the crystallizer was cooled by a cooling rate of 13.5~14.5 K/min. After the solution temperature in the crystallizer reached the crystallization temperature, the heat medium controlled at the crystallization temperature in the third water bath ⑦ was circulated to the jacket of the crystallizer. After that, the solution in crystallizer was held at a constant crystallization temperature. The time the solution in the crystallizer reached crystallization temperature, was defined as the starting point of induction time. The time interval between the time of appearance of crystals and the starting point was defined as the induction time. Also, the appearance of crystals in solution and the discrimination of the crystal form were observed by naked eye for about 230 minutes.

If the crystal form is observed in the experiment, the discrimination between α and β polymorphs of QE crystals appearing first in the solution can be immediately decided.

Concentration of primary nuclei

The polymorphs of QE appearing first in the supersaturated solution are shown in Figure 2 together with the solubility lines of α and β calculated from equations 1~6.

β occurred in a low supersaturated solution within the limits of temperatures between 305 and 311K, but the same crystal could not be observed at temperatures lower than 303K for induction time about 230 minutes. That is, the occurrence of α at temperatures lower than 303K, is independent of the concentration. On other hand, α occurred in a higher supersaturation at temperatures between 305 and 311K. In temperatures lower than 303K, the transformation rate of α to β by solution-mediated polymorphic transformation *(8)* seems to be very slow according to the results of this experiment and the results of composition changes of polymorphs in saturated solution previously reported *(2)*.

Induction time. The experimental data of the induction time θ of α and β were calculated with the equation 7 proposed by Harano et al *(6)* .

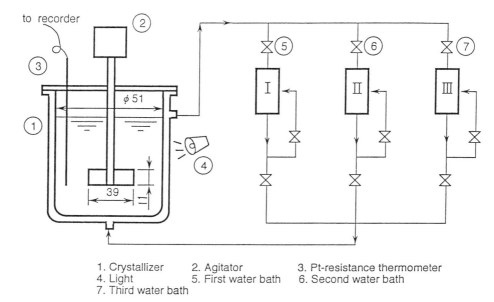

1. Crystallizer   2. Agitator   3. Pt-resistance thermometer
4. Light   5. First water bath   6. Second water bath
7. Third water bath

Figure 1. Schematic diagram of the experimental apparatus.

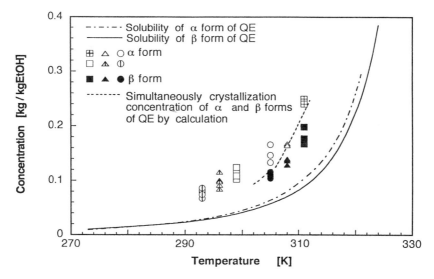

Figure 2. Polymorphs of QE occurred as primary nuclei.

$$\log\left(\theta \bullet V\right) = \log\left(\frac{N_{obs}}{k_N}\right) + \left\{\frac{16\,\pi v^2 \sigma^3}{3\left(2.303kT\right)^3}\right\}\left(\frac{1}{\log^2 S}\right) \tag{7}$$

where V [m^3] is the effective volume of crystallizer, k_N [$1/m^3 \bullet s$] the rate constant of nucleation, N_{obs} [-] the total number of nuclei at detecting time, k [J / K] Boltzmann's constant, v [m^3/molecule] the volume per molecule, σ [J / m^2] the surface energy, S [-] the supersaturation ratio = C/C_s ; C [kg-solute / kg-EtOH] the concentration of supersaturated solution, C_s [kg-solute / kg-EtOH] the concentration of saturated solution at a given temperature.

In the equation 7, if σ is independent of the supersaturation ratio, the relation between $\log(\theta \bullet V)$ and (1 / $\log^2 S$) is a straight line. Figure 3 shows the relation between $\log(\theta \bullet V)$ and (1 / $\log^2 S$) with experimental values of the induction time θ for α and β of QE with the crystallization temperature as the parameter. It is evident from Figure 3, that the experimental values are dispersed to some degree. However, it can be seen that the plots are approximately linear for each temperature and each crystal form. Values of (N_{obs} / k_N) and {$16\pi v^2\sigma^3/3(2.303kT)^3$} for each temperature and each polymorph in Figure 3 were calculated from intercept and slope of the straight lines, respectively.

Now, let us consider the (N_{obs} / k_N) term of the right side of equation 7. Values of (N_{obs} / k_N) for α and β QE have been plotted against the reciprocal of crystallization temperature in Figure 4. From the intercepts and slopes of the straight lines in Figure 4, the equations correlating (N_{obs} / k_N)α and (N_{obs} / k_N)β were obtained:

For α

$$\left(\frac{N_{obs}}{k_N}\right)_\alpha = 8.33 \times 10^{-19} \exp\left(\frac{1.45\times 10^{-19}}{kT}\right) \tag{8}$$

For β

$$\left(\frac{N_{obs}}{k_N}\right)_\beta = 3.21\times 10^{-18} \exp\left(\frac{1.45\times 10^{-19}}{kT}\right) \tag{9}$$

Since the straight lines for both crystal forms are parallel, we can conclude that the activation energies for diffusion are the same. Also, the frequency factors on an Arrhenius plot can be seen to have very small differences. These equations fit almost all of the experimental results within $\pm$ 5%.

On the other hand, the surface energy σ is considered in the {$16\pi v^2\sigma^3/3(2.303kT)^3$} term of the right side of equation 7. The slopes of the straight lines, SL, in Figure 3 are equal to {$16\pi v^2\sigma^3/3(2.303kT)^3$}. Therefore, σ can be calculated from the following equation:

$$\sigma = \left\{\frac{3\left(2.303kT\right)^3 SL}{16\,\pi v^2}\right\}^{1/3} \tag{10}$$

As is well known, the volume per molecule v, is given by the following relation:

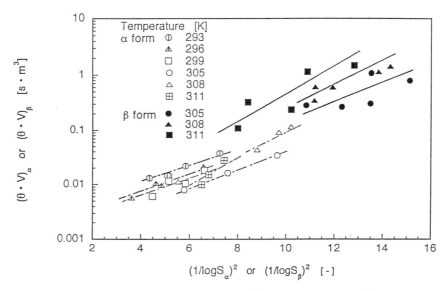

Figure 3. Relationship between $(\theta \cdot V)$ and $(1 / \log S_\beta)^2$.

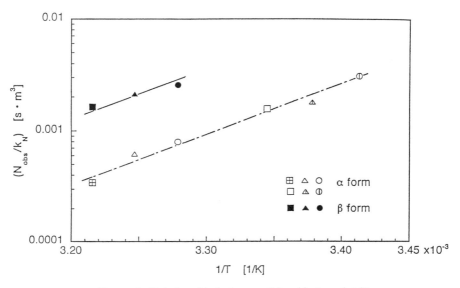

Figure 4. Relationship between (N_{obs} / k_N) and $1/T$.

$$v = \frac{M}{1000 \, \rho_s \, N_A} \tag{11}$$

where M [-] is the molecular weight, ρ_s [kg /m^3] the density of crystals, N_A[1/ mole] Avogadro's number. M and ρ_s of both crystal forms have the same vales 372.5 and 1360 kg/m^3 (2), respectively. Thus, the v value of both crystal forms becomes 4.55×10^{-28} m^3/ molecule.

In Figure 5, σ of α and β has been plotted against the crystallization temperature. The surface energy σ of both crystal forms is found to increase with increasing temperature in the range of 293 and 311K, and the surface energy of β is insignificantly larger than that of α in the range of 305 and 311K. These experimental results and the estimated values of the surface energy by the Nakai's equation (9) lie within the same order of the magnitude, 10^{-3} J / m^2. The curved lines of α and β in Figure 5 satisfy the following relationships at the crystallization temperature.

For α

$$\sigma_\alpha = -3.474 + 3.5219 \times 10^{-2} T - 1.1898 \times 10^{-4} T^2 + 1.3407 \times 10^{-7} T^3 \tag{12}$$

For β

$$\sigma_\beta = -10.4336 + 10.36183 \times 10^{-2} T - 3.43065 \times 10^{-4} T^2 + 3.7879 \times 10^{-7} T^3 \tag{13}$$

These fit almost all of the experimental results within 6%.

On the basis of equation 7, the experimental equations of induction time of QE from this work are given by the following relations.

For α

$$(\theta \cdot V)_\alpha = 8.33 \times 10^{-19} \exp\left(\frac{1.45 \times 10^{-19}}{kT}\right) \exp\left\{\frac{16 \, \pi v^2 \sigma_\alpha^3}{3(kT)^3 \ln^2 S_\alpha}\right\} \tag{14}$$

For β

$$(\theta \cdot V)_\beta = 3.21 \times 10^{-18} \exp\left(\frac{1.45 \times 10^{-19}}{kT}\right) \exp\left\{\frac{16 \, \pi v^2 \sigma_\beta^3}{3(kT)^3 \ln^2 S_\beta}\right\} \tag{15}$$

Equations 14 and 15 correlate induction time, crystallization temperature, supersaturation ratio, and surface energy. These equations fit almost all of the experimental results within $\pm 38\%$.

Polymorphs appearing first in a supersaturated solution

The type of polymorph appearing first from a supersaturated solution was considered

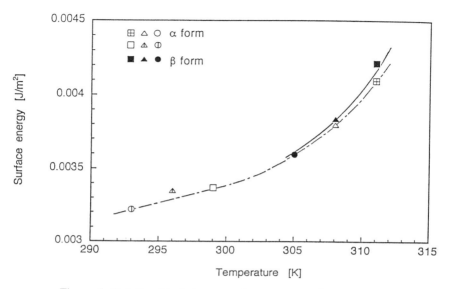

Figure 5. Relationship between surface energy and temperature.

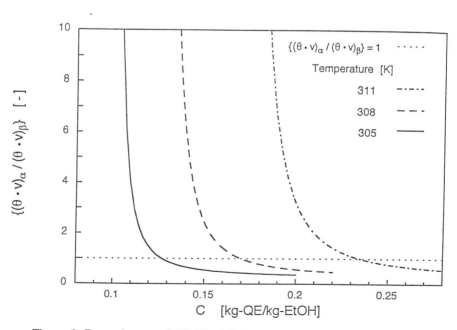

Figure 6. Dependences of $\{(\theta \cdot V)_\alpha / (\theta \cdot V)_\beta\}$ on the solution concentration.

from equations 14 and 15. In this connection, if equation 14 is divided by equation 15, the following equation is obtained.

$$\frac{(\theta \bullet V)_\alpha}{(\theta \bullet V)_\beta} = (0.2595) \ \frac{\exp\left\{\dfrac{16 \, \pi v^2 \sigma_\alpha^3}{3(kT)^3\left(\ln \dfrac{C}{C_{s\alpha}}\right)^2}\right\}}{\exp\left\{\dfrac{16 \, \pi v^2 \sigma_\beta^3}{3(kT)^3\left(\ln \dfrac{C}{C_{s\beta}}\right)^2}\right\}} \tag{16}$$

Equation 16 indicates that when $\{(\theta \bullet V)_\alpha / (\theta \bullet V)_\beta\}<1$, only α will occur first from a supersaturated solution.   Conversely, when $\{(\theta \bullet V)_\alpha / (\theta \bullet V)_\beta\}>1$, only β will occur first. Also, when $\{(\theta \bullet V)_\alpha / (\theta \bullet V)_\beta\}=1$, both forms of α and β will occur simultaneously .

The specific characteristic of equation 16 $\{(\theta \bullet V)_\alpha / (\theta \bullet V)_\beta\}$ was computed by the use of equations 2, 3, 5, 6, 12, 13, and 16 with crystallization temperatures of 305, 308, and 311K. The computed values of $\{(\theta \bullet V)_\alpha / (\theta \bullet V)_\beta\}$ are plotted against concentrations of supersaturated solution in Figure 6 with the temperature as a parameter. From Figure 6, as concentration of supersaturated solution increases, the value of $\{(\theta \bullet V)_\alpha / (\theta \bullet V)_\beta\}$ is smaller than 1 and only α will occur first.   Also, the concentration for simultaneous crystallization of α and β QE, $C_{\alpha\beta}$, when $\{(\theta \bullet V)_\alpha / (\theta \bullet V)_\beta\}=1$, are read from Figure 6 and are shown as a dotted line in Figure 2. As a result, $C_{\alpha\beta}$ is given by,

$$C_{\alpha\beta} \doteq 6.2 \times 10^{-15} \exp(0.1005 \, T) \tag{17}$$

This equation can be applied at temperatures between 305 and 311K.

On other hand, Figure 2 shows α occurring at temperatures lower than 303 K. This behavior can be considered in terms of equation 16. As the temperature decreases, σ_α becomes approximately equal to σ_β and $C_{s\alpha}$ becomes nearly equal to $C_{s\beta}$. As a result, $\{(\theta \bullet V)_\alpha / (\theta \bullet V)_\beta\}$ of equation 16 will be given by,

$$\{(\theta \bullet V)_\alpha / (\theta \bullet V)_\beta\} \doteq 0.26 \tag{18}$$

Therefore, the occurrence of α below 303K is possible.

Conclusions

1. Empirical equations for the solubility of the α and β forms of QE in ethanol are proposed.
2. The type of polymorph appearing first at the induction time could be correlated by a classical nucleation rate equation at temperatures from 293 to 311K.
3. The supersaturated ratio S, the surface energy σ, and the diffusion rate of the solute may considerably affect the crystal form appearing first in the supersaturated solution.

4. The concentration for the simultaneously crystallization of α and β QE, $C\alpha\beta$, in the range of 305 and 311K is given by:

$$C\alpha\beta \fallingdotseq 6.2 \times 10^{-15} \exp(0.1005\,T)$$

Acknowledgments

The quizalofop-ethyl used in this study was supplied by Nissan Chemical Industries, Ltd. I also wish to thank Nissan Chemical Industries, Ltd. for financial support.

References

1. Sakata, G.; Makino, K.; Morimoto, K.; Hasebe, S. *J. Pesticide Sci.* **1985,** *10* , 69.
2. Shiroishi, A.; Hashiba, I.; Kokubo, R.; Miyake, K.; Kawamura, Y. *Crystallization as a Separations Process;* Myerson, A. S.; Toyokura, K., Eds. ; ACS Symposium Series, Vol. 438.; American Chemical Society, Washington, DC, U.S.A., 1990; Chapter 19, pp 261–270.
3. Ura, Y.; Sakata, G.; Makino, K.; Kawamura, Y.; Ikai, I.*U.S. Patent, 4,629,493* **1986**
4. Ura, Y.; Hashiba, I . *Nippon Kagaku Kaishi* **1991,** *No.4,* 253.
5. Turnbull, D.; Fisher, J. C. *J. Chem. Phys.* **1949,** *17 ,* 19.
6. Harano, H .; Oota, K . *J. Chem. Eng. Japan* **1978,** *11 ,* 159.
7. Mitsuda, H.; Miyake, K.; Nakai, T. *Kagaku-Kougaku* **1967,** *31,* 1086. : *Int. Chem. Eng.* **1968,** *8,* 733.
8. Garti, N. *Crystallization and Polymorphism of Fats and Fatty Acids;* Garti, N.;Sato, K., Eds.;Surfactant Science Series, Vol. 31., Marcel Dekker Inc. : New York and Basel, U.S.A., 1988 ; pp 267-303.
9. Nakai, T. *Bull. Chem. Soc.* **1969,** *42,* 2143.

Chapter 11

Crystallization of Di-L-phenylalanine Sulfate Monohydrate from Fermentation Broth

Tetsuya Kawakita

Technology and Engineering Laboratory, Ajinomoto Company, Inc.,
1–1 Suzuki-cho, Kawasaki, Kanagawa 210, Japan

The characteristics of di-L-phenylalanine sulfate monohydrate crystals were investigated. Crystals of this compound were found to adopt two morphologies, i.e. needle form and thin plate form in the industrial plant. The former were metastable crystals and the latter were stable crystals which belonged to the same crystal system(monoclinic) as the pillar form crystals observed by powdered X-ray diffraction and solubility analyses. The needle form appeared at concentrations of L-phenylalanine higher than 30 g/dl at pH 1.7 to 2.0. On the other hand, the thin plate form appeared when the pH was below 1.0 ,or excess impurities (peak-A) derived from the mother liquid were accumulated in the crystallization liquid.

Di-L-phenylalanine sulfate monohydrate crystals($(C_9H_{11}NO_2)_2H_2SO_4H_2O$, Formula weight =446.5, crystal density =1.380 g/cm^3) were prepared by Matsuishi *et al.* .(2) to obtain the high purity product from the fermentation broth in industrial plant. Ordinarily this compound is composed of pillar form crystals and belongs to the monoclinic system(*1*),but changes in the crystal shapes to needle form or thin plate form were sometimes observed during the industrial operation of the crystallization process. The appearance of the particular shapes depend both on conditions during nucleation and on the subsequent growth of the crystal. In most industrial crystallization processes, several impurities are included in the operational solution, so the effects of impurities on the crystal shapes are considered to be practically important.
 In this paper, the morphological differences in the two crystal shapes are discussed with regard to the results of solubility and X-ray diffraction analyses and then the effects of various impurities on the change from the pillar form to thin plate form are studied.The substances used were selected to cover the major impurities included in L-phenylalanine solutions obtained from fermentation broth.

Back-ground

The industrial crystallization process examined is outlined in Figure 1.The filtered fermentation broth was fed to the vacuum evaporator and concentrated to 28 g/dl then the pH was adjusted within the range of 1.7 to 2.0 with sulfuric acid. When this solution was cooled continuously through a two-stage crystallizer whose temperature were 40 and 25°C, the pillar form crystals of the sulfate salt of L-phenylalanine formed and were easily separated by centrifugation.

When the solution was concentrated to more than 30 g/dl at the early stage of the operation, the needle form crystals became appeared and the viscosity of the slurry rapidly increased to more than 1,000 cp forcing the operation to stop.

On the other hand, when the large amount of the mother liquid of di-L-phenylalanine sulfate recycled into the filtered fermentation broth was increased to increase the recovery rate of L-phenylalanine. When these thin plate form crystals appeared, the crystal cakes were hardly separated from the slurry by centrifugation.

Experiment

Materials   Di-L-phenylalanine sulfate monohydrate in the pillar form : To 28 g/dl slurry of L-phenylalanine, 1.1 equivalents by volume of sulfuric acid were added with stirring at 60°C, and then cooled to room temperature with stirring. The pillar form crystals came out, the crystals were filtered then washed with water and dried in a vacuum desiccator with sulfuric acid.

Di-L-phenylalanine sulfate monohydrate both in the needle form and in the thin plate form : These two crystal cakes obtained from the industrial plant, separately, were washed with water and dried in a vacuum desiccator with sulfuric acid.

Pharmaceutical grade L-lysine monohydrochroride,L-arginine, L-citruline, L-valine L-leucine, L-isoleucine, L-tryptophan,L-tyrosine,DL-methionine,L-glutamic acid,L-asparticacid,L-alanine,glycine,L-threonine,L-serine,L-proline,L-glutamine,L-systeine, and sodium sulfate anhydrate were used.

Mother liquid of di-L-phenylalanine sulfate was sampled from the industrial plant.

Representative results of analysis of the mother liquid were as follows: L-phenylalanine 8.9 g/dl , sulfate ions 5.8 g/dl, nitrogen 2.4 g/dl, pH 1.6, Ai/phe[*] 3.6.

(*) Ai was the color value measured by the absorbence at 400 nm and phe was the content of L-phenylalanine measured by HPLC and detected at 210 nm (Type 638-50 ; Hitachi Co., Inc.; column CPK-08).

Characteristics of the two crystal shapes obtained from the plant
Difference in the crystal form   Each of the two different types of crystal , the needle form and the thin plate form, sampled from the industrial plant was analysed by a Geiger Flex powdered X-ray diffraction apparatus Type 2013 (Rikagaku kiki K.K.Japan) for the determination of the crystal form.

Measurement of solubility of the needle form and the pillar form crystals   For measurement, 30 ml of saturated sodium sulfate solution and 20 g of di-L-phenylalanine sulfate monohydrate either in the needle form or pillar form were added at 10 to 60°C, into 50 ml glass tubes with a spiral stirrer. Then, the contents were set in a water bath regulated with an accuracy of ±0.5°C for each experimental temperature. After stirring for about 24 hours, an aliquot was taken with a cotton-stopped pipette, and the concentration of L-phenylalanine in the filtrate was determined by HPLC (Type 638-50; Hitachi Co., Inc. Japan).

Velocity of transition between the needle form and pillar form   Measured amounts of the needle-shaped crystals were placed in 50 ml glass tubes with a spiral stirrer. To each tube was then added 40 ml of a saturated solution of the needle form at the experimental temperature and then the contents were stirred for a given period. The treated crystals were separated and dried in a vacuum desiccator. The amount of the needle form was measured by the X-ray diffraction method. The operational temperatures were 30, 40 or 60°C. Twenty or 60 % of the pillar form crystals were added to the saturated solution of the needle form as the saturating body. The velocity of the transition of the needle form into pillar form crystals was calculated according the following equation:

$$V=(1-X_t / X_0)/t$$

where V is the velocity of the transition (1/hr), X_t is the contents of the needle form in sampled crystals at a time t (%), X_0 is the contents of the needle form in initial crystals sampled from the glass tube (%) and t is the elapsed time (hr).

Changes in crystal shape from the pillar form to the thin plate form

Apparatus and procedures for crystallization   The slurry solution of di-L-phenylalanine sulfate (28 g/dl as L-phenylalanine) was prepared in the capacity of 1.0 liter of the three-necked glass crystallization vessel set in a water bath regulated at a temperature of 60°C and was agitated. When the solution became transparent, it was allowed to cool and then the temperature reached 40°C, measured amounts of the pillar form crystals prepared previously were added as seeds. The slurry was continuously agitated and cooled to 25°C at a rate of about 5°C/hr. The slurry was then sampled with a spoite and crystal photographs were taken using an optical microscope (50× magnification). The length, the width and the thickness expressed by the c-axis, the b-axis and the a-axis directions of the crystal, respectively, were measured with Nikkon Shadow-Graph Model 6. We also investigated the effects of impurities , sodium sulfate, pH, amino acids and the mother liquor on the change in crystal shape.
Sodium sulfate was added to the solution at 50 and 100 wt% relative to L-phenylalanine. Amino acids were added at 2,3 and 7 wt% relative to L-phenylalanine. The mother liquor, which was decolorized previously by a given amount of activated carbon was added to the solution in the range from 20 to 58 wt% relative to L-phenylalanine and the pH was adjusted to 1.7 with sulfuric acid. To determine the effects of pH, the value was adjusted to 0.4, 0.8 and 1.7 with sulfuric acid.

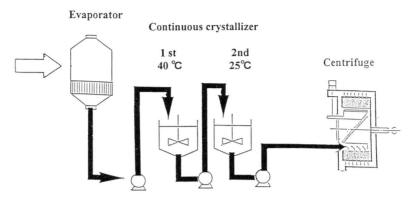

Figure 1. Flow diagram of crystallization process in the industrial plant

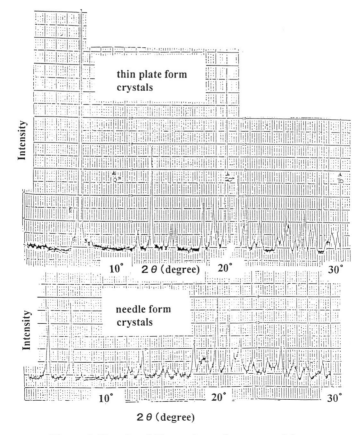

Figure 2. Powdered X-ray chart of two crystal forms

Chromatographic analysis of mother liquid adhering to crystals   The pillar form crystals or the thin plate form   obtained from the industrial plant were separately dissolved in water. For protein analysis, these two dissolved samples were fed into the chromatograph column -PROTEIN I-125 (Water's Co., Inc.) which was preconditioned with 0.05 M phosphate buffer solution (pH 6.8) and then developed by 0.05M Na_2HPO_4-Na_2PO_4 buffer solution (pH6.8). The eluate was measured by absorption at 280 nm. For comparison, mother liquids of either crystal forms separated in the industrial plant were also analyzed.

Results and Discussion

Characteristics of the two forms of crystals   The crystal appearances and other characteristics of the two types of crystals obtained from the industrial plant are summarized in Table 1. The gel-like highly viscous slurry obtained from the pipe line contained more than 30 g/dl of L-phenylalanine as needle form crystals. On the other hand, the slurry obtained from the outlet of the centrifuge consisted of thin plate form crystals as described above. These two types of crystals were different from the ordinary pillar form crystals. Figure 2 shows the powdered X-ray diffraction chart of these two forms of crystals. The chart of the thin plate form crystals had the same pattern as the pillar form crystals reported by Matsuishi(*2*).

On the other hand the chart of the needle form crystals showed a different peak from the pillar form crystals. The solubility data of di-L-phenylalanine sulfate monohydrate and sodium sulfate are shown in Figure 3. These experimental results show that the needle form crystals is more soluble than the pillar form crystals.

Thus, the needle form crystals are the meta-stable form and the pillar form crystals are the stable form in solution.

Transition from the needle form to the pillar form   In the solution containing more than 30 g/dl of di-L-phenylalanine sulfate at 60 °C , needle form crystals were often observed with cooling in the laboratory. These conditions are similar to those in the industrial plant under which formation of needle form crystals has been observed. Needle form crystals are formed at   concentrations of di-L-phenylalanine sulfate in excess of 30 g/dl at pH 1.7 to 2.0 . The transition time required from the needle form crystals obtained from the industrial plant to pillar form was 22 hours at 60°C even when the pillar form crystals which amounted to 50% of the solute in a saturated solution was added. When no seed was added, the needle form crystals remained unchanged for more than 80 hours at 60°C.

Assuming that the velocity of the transition of the needle form is caused by the dissolution of the needle form crystals and crystallization of the pillar form due to the difference in the solubility of the two crystal forms, the velocity of the transition might be a function of the degree of supersaturation of the pillar form, and will be presented by the following equation(*3*):

$$V = k \Delta C$$

Table 1. Characteristics of two crystals obtained from the industrial plant

Crystal Appearance	needle form	thin plate form
photograph 1000 μ ⌐ 0 ⌐		
Final concentration of L-phenylalanine in evaporator (g/dl)	> 30	25 ~28
mother liquid viscosity (cp)	>1,000	20 ~30

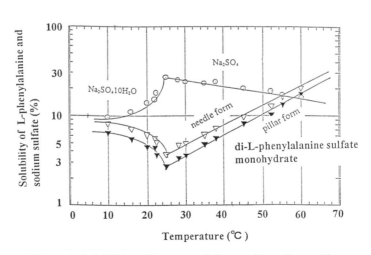

Figure 3. Solubilities of two crystal forms with sodium sulfate

where V is the velocity of the transition (1/hr), $\Delta C(=C-C_0)$ is the difference between the concentration of di-L-phenylalanine sulfate in the solution(C: g/dl) and the solubility of the pillar form crystals at temperature(C_0 :g/dl) and k is a constant.

The k values obtained from the experiments in which pillar form crystals were added at 50% [*] were as follows; 2.12×10^{-2} at 60°C, 1.78×10^{-2} at 50°C, 1.47×10^{-2} at 35°C.

(*) 50% means that 50% of the solute dissolved in a saturated solution.

The transition enthalpy ΔH was calculated by applying these value of k constants to the Arrhenius equation:

$$k=A\exp(-\Delta H/RT)$$

The transition enthalpy required was 12.2 kJ/mol. The retardation of transition of the needle form crystals sampled from the industrial plant may be the effect of unknown impurities.

Change in crystal appearance of the pillar form to the thin plate form

Effects of sodium sulfate   The thickness of the thin plate form crystals in the a-axis direction is very thin compared with that of ordinary pillar form crystals. The thickness in the a-axis and the lengths in other axes relative to the a-axis against the concentration ratio of sodium sulfate to L-phenylalanine in the solution are summarized in Table 2. The thickness in the a-axis showed a constant value of about 50μ m irrespective of changes in the concentration of sodium sulfate. The length in the b-axis relative to the a-axis remained unchanged at 1.4 in the presence of sodium sulfate existed. On the other hand, the   length in the c-axis relative to the a-axis decreased markedly from 25 to 6.5 with an increase in the ratio of sodium sulfate to L-phenylalanine in solution. As the   ratio of sodium sulfate to L-phenylalanine concentration increased, the growth of crystals in the c-axis was suppressed and   the crystal shape changed from pillars to fine rods. However, sodium sulfate itself was not responsible for the change to the thin plate form.

Effects of the pH   The dependencies of the length in the b-axis and the lengths of other axes relative to that in the b-axis on the pH are summarized in Table 3. The length in the b-axis was unchanged within the range of the pH tested. The length in the a-axis relative to the b-axis decreased with decreases in pH of the solution. On the other hand, the lengths of the c-axis relative to the b-axis was in the range of 2.2 to 3.6. It follows that the growth in the a-axis direction was suppressed by hydrogen ions and with increases in the pH the crystals adopt the thin plate form.

Effects of amino acids   The effects of 18 kinds of amino acids on crystal appearance are summarized in table 4. Among these amino acids tested, only leucine and tryptophan affected the change in crystal form from pillars to thin plates at concentrations relative to L-phenylalanine of more than 3%. These two amino acids are hydrophobic, so they might interact with the L-phenylalanine skeleton in the crystal structure of di-L-phenylalanine sulfate monohydrate and are supposed to suppress growth in the a-axis direction. Isoleucine, valine and tyrosine which are analogous

Table 2. Effects of sodium sulfate on the change in crystal length in the a-axis and relative lengths in other axes

operational condition
    pH in solution : 1.7
    Crystallization : Temperature change from 40℃
           to 25℃ for 3 hours

Sodium sulfate/phe (g/g in solution)	length in a-axis (μ)	relative length	
		b/a	c/a
0	50	3.6	25.2
0.5	45	1.45	21.1
1.0	52	1.44	6.5

Table 3. Effects of pH changes on crystal length in the b-axis and relative lengths in other axes

operational condition
    Content of sodium sulfate : 50% to L-phenylalanine
           in solution
    Crystallization : Temperature change from 40℃
           to 25℃ for 3 hours

pH	length in b-axis (μ)	relative length	
		a/b	c/b
0.4	104	0.038	2.23
0.8	104	0.086	3.61
1.7	112	0.446	2.73

hydrophobic amino acids did not influence the change in the crystal shape. It is thus necessary to investigate the effects of amino acids on changes in the crystal shape, in more details. In phenylalanine fermentation broth, amino acids are present at ratios of less than 3% to L-phenylalanine and impurities contained in the fermentation broth other than these amino acids might affect the change to the thin plate form.

Effects of the mother liquid on the change in the crystal appearance The effects of the mother liquid decolorized by activated carbon on the change in crystal appearance are summarized in Table.When less than 1.05 ml of the mother liquid was added to L-phenylalanine in solution, only pillar form crystals were obtained. When 1.2 ml of the mother liquid was added, the thin plate form crystals began to appear. At 2.5 ml of the mother liquid, only thin plate form crystals were obtained irrespective of increases in the amount of activated carbon used for decolorization. Furthermore, when activated carbon was added at a ratio of 1.6 % to L-phenylalanine, which is about 8 times as much as that used in the industrial process, and thin plate form crystals appeared although the color value of Ai/phe was diminished to 0.08 which almost corresponded to a transparent solution.

This result indicates that colored substances and factors which were adsorbed onto activated carbon did not affect the change into thin plate form crystals. The figures in parentheses in Table 5 show the percentages of the mother liquid content included in crystal cakes. When Mc(-) is the mother liquid content in crystal cakes and Ms (ml/g) is the mother liquid added to crystallization solution, Mc was proportional to $Ms^{0.5}$ as expressed by the following empirical equation :

$$Mc = 0.368 Ms^{0.5}$$

The frequency of the appearance of thin plate form crystals was dependent almost entirely on the mother liquid content in crystal cakes and was independent of the degree of removal of colored substances which were adsorbed onto the activated carbon.

The elution chart for impurities of crystals and mother liquid obtained from a Protein Column I-125 are presented in Figure 4. There was no difference other than peak A between the pillar form and the thin plate form crystals. Although the amount of peak A detected in the mother liquid of the thin plate form crystals was almost the same as that in the mother liquid of the pillar form crystals, peak A was detected only in the thin plate form crystals. The location of peak A was close to that of L-phenylalanine, so this compound must have characteristics similar to those of L-phenylalanine. Thus, peak A might be one of the factors which affect the change from pillar form to thin plate form crystals. Future studies should be performed to identify the structure of peak A substance and clarify the influence of this substance on the change in the crystal appearance.

Conclusion

1) Two morphologically, different forms of di-L-phenylalanine sulfate monohydrate crystals were obtained in the industrial plant ; i.e. needle form and thin plate form. The former was metastable and the latter was stable. The needle form crystals appeared at concentrations of L-phenylalanine higher than 30 g/dl within the pH range of 1.7 to

Table 4. Effects of various amino acids on the changes in appearance of crystals

legend

Frequency of the thin plate form appeared
++	more than 50%
+	10 ~ 50%
±	less than 10 %
—	no effect

amino acid/L-phe (%)	Lys	Arg	Cit	Val	Leu	Ileu	Trp	Tyr	Met
7	—	—	—	—	++	—	++	—	—
3	—	—	—	—	±	—	+	—	—
2	—	—	—	—	—	—	—	—	—

amino acid/L-phe (%)	Glu	Asp	Ala	Gly	The	Ser	Pro	Gln	Cys
7	—	—	—	—	—	—	—	—	—
3	—	—	—	—	—	—	—	—	—
2	—	—	—	—	—	—	—	—	—

Table 5. Effects of the mother liquid on crystal appearance

Amount of mother liquid added (ml/g-L-phe in sol.)	Amount of activated carbon added (% to L-phe in solution)				
	0	0.2	0.4	0.8	1.6
0.44 (20)	◎ 0.27				
0.90 (34)	◎ 0.29	◎0.26			
1.05 (37)	◎ 0.27			◎0.13	
1.2 (40)	◎△ 0.47	◎△0.28	◎△0.24		◎△0.08
1.4 (44)	◎△ 0.34			◎△ 0.18	
2.5 (58)	△ 0.40			△ 0.24	

◎   pillar form crystals   △   thin plate form crystals

Figures in parenthes in left column represent the percentages of the mother liquid content in crystal cakes.

Figures in the Table represent the values of Ai/L-phe in the crystals.

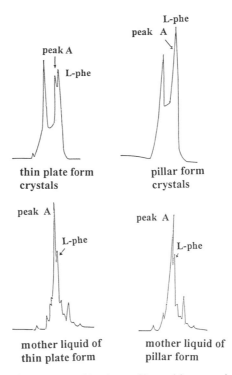

Figure 4. Chromatographic chart of impurities contained in the crystals and their mother liquids

2.0.   The needle form crystals obtained from the industrial plant showed hardly any transformation to the pillar form without seed. When more than 20% of the pillar form crystals was added as seed, the transition was observed at higher temperatures. The transition enthalpy was estimated 12.2 kJ/mol.

2) Thin plate form crystals belonged to the same crystal form as pillar form crystals and was obtained by the suppressed growth in the direction of the a-axis. This crystal form usually appeared under conditions of (a) pH less than 1 and (b) inclusion of the mother liquid at more than 40% in crystal cakes.

The key substances   governing the change into the thin plate form were investigated. The two amino acids leucine and tryptophan were effective when they present at a ratio of 3% relative to L-phenylalanine in solution.The peak A was separated from the thin plate form crystals by chromatography using Protein I-125(Water's Co. Inc.). The compound in peak A might be the factor responsible for the observed change in the crystal appearance.

Acknowledgment

The author wishes to express his sincere thanks to Mr. K Iitani, Manager of Technology & Engineering Laboratories of Ajinomoto Co., Inc. for   permission to submit this paper and to Mr. Koga and Mr. Igarashi for their help in this experiment at the Kyushu Ajinomoto Co., Inc plant.

References

(1) Nagashima, N; Sano,C; Kawakita,T; Iitaka I; Analytical Science,**1992**,8,723-725
(2)Matsuishi,T;   Kabashima,A; Murayama,H; Kitahara,T;   Toyomasu,R.,Japan Patent,**1981**,56-79652; Chem.Abstr. ,**1981**,95,187674Z.
(3) Sakata,Y ; Agricultural and Biochemical Chemistry,**1962**,26(6),355-361

Chapter 12

Characterization of Aluminum Trihydroxide Crystals Precipitated from Caustic Solutions

Mei-yin Lee[1], Gordon M. Parkinson[2], Peter G. Smith[3], Frank J. Lincoln[1], and Manijeh M. Reyhani[2]

[1]Research Centre for Advanced Mineral and Materials Processing, Department of Chemistry, University of Western Australia, Nedlands, Western Australia
[2]A. J. Parker Cooperative Research Centre for Hydrometallurgy, School of Applied Chemistry, Curtin University of Technology, GPO Box U 1987, Perth 6001, Western Australia
[3]A. J. Parker Cooperative Research Centre for Hydrometallurgy, Division of Minerals, CSIRO, P.O. Box 90, Bentley, Western Australia

Gibbsite is one of the polymorphs of aluminium trihydroxide and is produced commercially by the Bayer Process through its crystallization from sodium aluminate solutions. This work demonstrates that there is a complex interplay of factors which affect the polymorphism and morphology of aluminium trihydroxide crystals precipitated from concentrated aluminate solutions. The formation of gibbsite is favoured at higher temperatures, while bayerite is formed predominantly at room temperature. The nature of the alkali metal ion present in the aluminate solution has a substantial influence on the morphology of the single crystals of gibbsite formed, with hexagonal plates resulting from sodium aluminate solutions and elongated hexagonal prisms from potassium aluminate solutions.

The conversion of bauxite ore into alumina, via the Bayer process, is a well established major industry. However, the rate of precipitation of aluminium trihydroxide as gibbsite is an extremely slow process and is not well understood. Moreover, difficulties are encountered in regulating particle growth to achieve desirable characteristics of purity, strength, morphology and size distribution. As part of an overall programme to understand the mechanism and the rate of crystal growth from solution, the process of crystal growth on different faces of synthetic gibbsite is being studied by *in-situ* and *ex-situ* microscope techniques. The work reported here concentrates on the initial characterization of different of aluminium trihydroxide crystals produced by precipitation from aluminate solutions.

There are three polymorphs of aluminium trihydroxide: gibbsite, bayerite and nordstrandite. The difference between the three polymorphs is in the stacking order of their $[Al_2(OH)_6]_n$ double layers, and the inclusion of impurities may affect this stacking sequence.[1] Bayerite has been reported to form as somatoids, with an "hour glass", cone or spindle shape (somatoids are defined to be "bodies" of uniform shape that are not enclosed by crystal faces), while gibbsite forms as agglomerates of tabular and hexagonal shaped crystals (1).

A study by Misra and White (2) showed that single crystals of up to 40μm can be produced from potassium aluminate solutions (rather than sodium aluminate solutions) under conventional Bayer process type conditions. Wefers (3) suggested that the inclusion of potassium ions results in the formation of elongated crystals that are morphologically quite perfect, whilst with the inclusion of sodium ions, less well formed crystals with growth distortions result. It is believed these cations are incorporated by substituting for the hydrogen atom of a hydroxyl group, and as a result of the different sites of the cation incorporation, different crystal morphologies can result (4).

Carbonation of potassium aluminate solutions at elevated temperatures (70-80°C) is reported to result in the formation of large (~80 μm) single crystals of gibbsite (Rosenberg, S.P. private communication, 1993). A study by Wojcik and Pyzalski (5) however, showed bayerite to be the predominant species formed in the carbonation process of potassium aluminate and sodium aluminate solutions between 50 - 80°C.

In this work, several methods of growing crystals were followed, using both sodium aluminate and potassium aluminate solutions, at either room temperature or 70°C and the precipitates were characterised using scanning electron microscopy to examine the morphology of the crystals, and powder X-ray diffraction to identify the major phases present.

Experimental

Liquor Preparation. The aluminate solutions (synthetic Bayer liquors) were prepared using gibbsite (C31; Alcoa Chemical Division, Arkansas), sodium hydroxide pellets (AR Grade) and sodium carbonate (AR Grade) for sodium aluminate; or with potassium hydroxide (AR Grade) and potassium carbonate (AR Grade) for potassium aluminate solutions.

A mixture of gibbsite, caustic and deionised water was heated, with stirring, in a stainless steel vessel until all the gibbsite had dissolved. This was then added to pre-dissolved carbonate. The solutions were allowed to cool and made to volume with deionised water to obtain a final concentration of 140g/L Al_2O_3, 200g/L total caustic (expressed as Na_2CO_3) and 240g/L total alkali (expressed as Na_2CO_3), in line with the North American alumina industry terminology.

The solutions were filtered through a 0.45μm membrane prior to use.

Solid Preparation. The precipitate from each experiment was collected by filtration through a 0.45μm membrane, washed with hot de-ionised water and air dried at room temperature. The precipitates were characterised using powder X-ray diffraction and scanning electron microscopy.

Precipitation Conditions. The amount of aluminium dissolved in the synthetic Bayer liquors studied here is in excess of the solubility of the $Al(OH)_3$ polymorphs, and solid will precipitate out if the solutions are simply left for a sufficiently long period of time, referred to as ageing (4 - 48 hours, depending on conditions). The ageing period included the crystallization induction time and some precipitation, but the rate of precipitation can be increased by partially neutralising the alkali hydroxide by the addition of carbon dioxide or acid, or by the addition of seed crystals.

Ageing of Aluminate Solution. The aluminate solutions were allowed to age for 48 hours at room temperature or for 4 hours at $70°C$ in the absence of any seed crystals or the addition of acid or CO_2.

Addition of Carbon Dioxide Gas (Carbonation). Bubbling carbon dioxide through a tube into synthetic Bayer liquors at moderate rates results in the exit holes for the gas becoming blocked by precipitated gibbsite. To try to overcome this problem, and hence achieve reproducible rates of precipitation, two approaches were adopted. First, carbon dioxide was bubbled through a bubbler ring with many holes into a stirred sodium or potassium aluminate solution at room temperature or $70°C$. Secondly, carbon dioxide was bubbled through a gas chromatography syringe (0.5mm) into a potassium aluminate solution at $70°C$, with no stirring.

Addition of Acid. Hydrochloric acid (0.5moles/L) was added dropwise via a peristaltic pump to a stirred sodium or potassium aluminate solution at room temperature or $70°C$, at a rate of 0.5mL/min. The volume of acid added completely neutralised the caustic solution.

Addition of Seed. Sodium or potassium aluminate solutions were seeded with 10g/L of gibbsite (C31) in Nalgene bottles and tumbled end over end at $70°C$ for 2 hours.

Results

For all conditions, bayerite was the predominant species formed at room temperature, while the formation of gibbsite was favoured at $70°C$.

Bayerite Formation.

Ageing of Aluminate Solution. The initial products from ageing sodium aluminate solution at room temperature (after 16 hours) were mainly calcium containing compounds ($CaCO_3$ and $3CaO.Al_2O_3.CaCO_3.11H_2O$) and $FeO(OH)$, and some bayerite. Calcium and iron are present as impurities in the reagents used for liquor preparation. The morphology of these precipitates consisted of rounded aggregates of plates. Bayerite with a "woolball" morphology was observed after 48 hours (Figure 1).

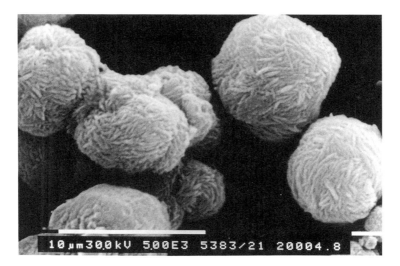

Figure 1: Bayerite with "woolball" morphology from agein sodium aluminate solutions

Figure 2: Bayerite from ageing potassium aluminate solutions

The morphology of bayerite formed by ageing potassium aluminate solutions consisted of agglomerates of hexagons and "diamonds' as well as single crystals of elongated hexagonal prisms (Figure 2) These elongated prisms are typical of the morphology of crystals grown from potassium aluminate solutions.

Addition of Carbon Dioxide Gas (Carbonation). The bayerite crystals produced by the carbonation of sodium aluminate consisted of "web-like" and "frond" shaped crystals (Figure 3), while from potassium aluminate solutions, bayerite consisted of agglomerates of fine crystals, with some "web-like" and "frond" shaped crystals. These were quite unlike the somatoid shapes described in the literature.[1]

Addition of Acid. The addition of acid (HCl) to the aluminate solutions at room temperature resulted in the formation of two distinct morphologies of bayerite. In both sodium aluminate and potassium aluminate, the bayerite consisted of agglomerates of elongated triangular prisms that showed radial growth characteristics (Figure 4). From potassium aluminate solutions, bayerite also took the form of single crystals (~40 mm), and sometimes with "zig-zag' edges (Figure 5).

Gibbsite Formation.

Ageing of Aluminate Solution.   Ageing both sodium aluminate and potassium aluminate solutions at 70°C resulted in the formation of gibbsite and nordstrandite.   Trace amounts of bayerite were also observed in aged sodium aluminate solutions. The morphology of gibbsite formed by ageing aluminate solutions is similar to that of natural gibbsite. Agglomerates of hexagonal tablets as well as "diamond" shaped crystals were observed from both sodium aluminate and potassium aluminate solutions, with the finer crystals from sodium aluminate solutions. Single crystals were also formed in both cases, elongated hexagonal prisms from potassium aluminate and hexagonal tablets from sodium aluminate solutions. Figure 6 is a micrograph of the crystals collected from potassium aluminate solutions.

Addition of Carbon Dioxide Gas (Carbonation). Addition of carbon dioxide through a bubbler ring into a stirred aluminate solution at 70°C resulted in the formation of gibbsite and bayerite, while carbonation of potassium aluminate though a single inlet resulted in the formation of relatively large, single crystals of gibbsite (Figure 7). The morphology of the bayerite/gibbsite crystals consisted of "web-like" structures with protruding hexagonal prisms (Figure 8).

Addition of Acid. The crystals formed by acid addition to aluminate solution were agglomerates of hexagons, diamonds and elongated prisms.   Again, coarser crystals resulted from potassium aluminate solutions (Figure 9).

Figure 3: Bayerite from carbonation of sodium aluminate solutions

Figure 4: Bayerite from acid addition to potassium aluminate

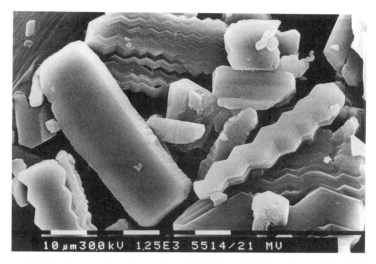

Figure 5: Single crystals of bayerite with "zig-zag" edges from acid addition to potassium aluminate solutions

Figure 6: Gibbsite from ageing potassium aluminate solution

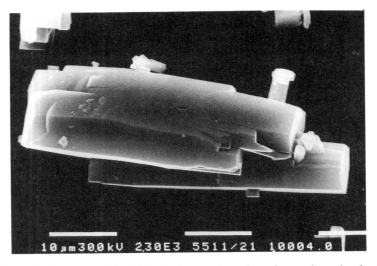

Figure 7: Single crystal of gibbsite from carbonation of potassium aluminate solutions

Figure 8: Gibbsite/bayerite from carbonation of sodium aluminate

Figure 9: Gibbsite from acid addition to potassium aluminate solutions

Figure 10: Industrially produced gibbsite (C31)

Figure 11: Gibbsite from seeding potassium aluminate solutions

Addition of Seed. Seeding sodium aluminate solutions with gibbsite (C31) at 70°C resulted in agglomerates of hexagons and diamonds, as well as single crystals of hexagonal plates. The morphology of these crystals is similar to that of industrially produced gibbsite (Figure 10). Seeding potassium aluminate solutions with gibbsite (C31) resulted in the formation of crystals typical for this system - elongated hexagonal prisms of gibbsite (Figure 11).

Conclusions

- Bayerite formation is favoured at room temperature while gibbsite formation is favoured at 70°C.

- The morphologies of bayerite are not necessarily somatoids and will depend on the method of preparation.

- Although the morphology of gibbsite varies with the crystallization conditions, the crystals are generally agglomerates of hexagons and/or diamonds or single crystals of hexagonal plates from sodium aluminate, and single crystals of elongated hexagonal prisms from potassium aluminate solutions.

- The different morphologies of gibbsite crystals grown from sodium and potassium aluminate solutions, under otherwise identical conditions, suggest that the alkali metal plays an important role in the crystallization mechanism.

Acknowledgments

This work has been supported under the Australian Government's Cooperative Research Centres programme, and by the Australian Mineral Industries Research Association and this support is gratefully acknowledged. The support of the Centre for Microscopy and Microanalyses, University of Western Australia is also acknowledged.

References

1. Wefers, K.; Misra, C. *Alcoa Technical Paper No 19*, **1987** Revised
2. Misra, C.; White, E.T. *J Crystal Growth*, **1971,** *8*, 172
3. Wefers, K. *Die Naturwissenschaften*, **1962,** *49*, 1
4. Lee, M.; Rohl, A.L.; Gale, J.D.; Parkinson, G.M.; Lincoln, F.J. Trans IChemE, **1996,** *74A*, 739
5. Wojcik, M.; Pyzalski, M. *Light Metals*, **1990,** 161

STUDIES RELATED TO INDUSTRIAL CRYSTALLIZERS AND PROCESSES

Chapter 13

Features of the High-Pressure Crystallization Process in Industrial Use

M. Moritoki[1,2], N. Nishiguchi[1], and S. Nishida[1]

[1]Engineering and Machinery Division, Kobe Steel, Ltd.,
3–18 Wakinohamacho 1-Chome, Chuoku, Kobe 651, Japan

High pressure crystallization is a separation technology characterized by the application of very high pressure to a mixture in order to crystallize a component in it, followed by the separation of the mother liquor under pressure, and by the further separation o,f residual mother liquor together with a small part of the crystals that just melted during the pressure release. As the pressure plays the role of driving parameter not only for crystallization, but also for separation and purification, it should be said that this crystallization is quite different from the conventional crystallization by cooling. In fact, as a result of investigation and practical use, it was found that this crystallization has a lot of unique features caused by the nature and functions of pressure. This paper deals with these features of the high pressure crystallization from the view point of industrial use, in relation to the nature and functions of pressure.

It is well known that liquid can be solidified by the application of pressure as well as by cooling. It was thought that this phenomenon could be used for crystallization. The first primitive experiments were carried out with benzene colored with methyl-red in light red. The first separated liquor under the pressure was like black, rather than blood red. Then the second became red, and the third was light yellow. The color lightened step by step, and finally the separated sample was almost transparent. Thus, the high pressure crystallization experiments started.

In experiments, high pressure vessels of twenty or thirty milli-liters in capacity were used, and pressure was applied gradually so as to prevent the increase of the temperature of the sample. And then, liquid was separated. Most of our fundamental data were obtained with this method.

[2]Current address: 11–6 Higashi 3-Chome, Midorigaoka-cho, Mikishi, Hyogo 673–05, Japan

For industrial use, however, an adiabatic process with a short cycle time was required. Owing to the nature of pressure, it was found that the temperature inside the high pressure vessel is always uniform during an adiabatic pressure application if the initial state is uniform. Besides, it was also found that the growth rates of crystals are extremely fast in many compound systems so that they reach a new equilibrium state in a few seconds after the adiabatic pressure change. If the separation of the mother liquor was carried out after the pressure application within the time in which the temperature remained uniform, there was no difference between the state of isothermal and adiabatic procedures except for temperature.

The temperature after the pressurization could be controlled by adjusting the feed temperature.   As a results, experimental data obtained isothermally with a small vessel can be applied to process design for adiabatic use in industrial scale.

The above considerations were verified by the actual operation of a pilot plant and separation plant for p-cresol. In addition, many unexpected features caused by the nature of pressure were newly discovered. At the same time, through experiments with a small apparatus, new fields of crystallization carried out only under high pressures have been found.

In this paper, the concept of an adiabatic procedure will be introduced first together with the features caused by the nature of pressure. Then, examples of the newly discovered fields of application will also be briefly examined along with possibilities for future use.

Concept of the Adiabatic Process

Figure 1 schematically shows the temperature change of a eutectic mixture during an adiabatic pressure cycle (*1*). The lines X_1, X_{10} and X_{1e} show the melting lines of the component 1, the feed mixture and the mixture of eutectic concentration, respectively. The line X_{11} shows the melting line of the mother liquor to be separated.

When pressure is applied to the feed mixture in a high pressure vessel at Point A, which is higher than the melting temperature, compression heat increases the temperature of the feed slightly, and at Point B, crystallization starts. As the pressure rises, the amount of crystals increases accompanied by a large temperature rise caused by the latent heat. Then, the concentration of the component in the liquid phase decreases to X_{11} as shown by Point C, where the mother liquor is separated from the solid phase, or removed from the vessel.

Before feeding, the feed temperature may be decreased to a temperature, as shown by Point A', lower than the melting point of the feed so as to obtain a slurry state. This reduces the separation pressure as shown by Point C' in Figure 1. If the mother liquor is thoroughly removed from the system after the crystallization, the pure component 1 will be obtained and its fraction to the feed will be $(X_{10}-X_{11})/(1-X_{11})$.

In practice, however, some amount of mother liquor remains between crystals. From this state, when the liquid pressure is decreased slightly, a small amount of crystals melts at their surfaces so as to dilute impurities in the remaining liquor. And then the diluted liquor is further separated. Such a partial melt and dilution is called sweating. The sweating and the liquid removal can be carried out repeatedly or continuously towards atmospheric pressure. As a result, the impurity concentration decreases exponentially.

The sweating due to the pressure decrease causes a small temperature decrease due to the latent heat. This temperature decrease prevents an excess amount of partial melt from occurring. This step is also schematically shown by the broken lines in Figure 1.

During the sweating by the liquid pressure decrease, the mean pressure applied to the remains in the vessel can be kept higher than with the liquid, such as that applied at the beginning of separation. The concept is shown in Figure 2. This pressure difference can be utilized for the compaction of crystals, that is, for decrease of the liquid fraction and improvement of the purity of the remains in the vessel.

It was found that even if the crystals were compacted and the liquid fraction was considerably reduced, the paths of liquid removal were maintained as pinholes through the inside of the cake of crystals, as long as the liquid remained impure. These pinholes vanish at the end of separation.

The sweating and the solid compaction progress in parallel. This is one of the important reasons why high purity is easily obtained with a high yield rate in high pressure crystallization.

Flow System for Industrial Uses

A typical flow diagram for the adiabatic process is schematically shown in Figure 3(2). The feed is first adjusted to a temperature previously determined, and sent to a high pressure vessel. Pressure is then applied by a piston driven by a hydraulic oil unit. After the pressure reaches a predetermined value, a valve in the outlet line is opened and the mother liquor flows out. In the outlet line, a nozzle with a very small diameter is fabricated so as to regulate the flow rate of the mother liquor. As the oil pressure can be controlled as a constant, the piston is inserted inside the vessel according to the mother liquor removal so as to keep the piston head pressure as a constant. And crystals are compacted with liquid pressure decrease, which accompanies the sweating. Finally, the purified and compacted crystals form a bulky solid or cake.

After the separation, the side cylinder of the high pressure vessel in the crystallizer is lifted to remove the cake. Then it is closed again to start the next cycle. The whole process is run repeatedly controlled by a computer.

Feature of Industrial Use

As described above, the pressure is the only parameter in this process. The energy of pressure propagates at the sound velocity. Mass transfer caused by pressure difference is also very fast. Therefore, the pressure always works uniformly in the liquid phase throughout the vessel.

In addition, as the changes in volume due to compression and liquid-solid transition are substantially very small, the pressure can be controlled very quickly and easily.

Those two fundamental properties of pressure cause the various features of high pressure crystallization. These features can be described as follows.

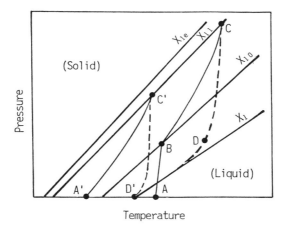

Figure 1. Concept showing temperature change during an adiabatic pressure increase and separation, and then sweating.

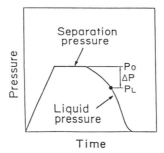

Figure 2. Change in the liquid pressure P_L during a cycle, keeping the inside pressure P_0 constant.

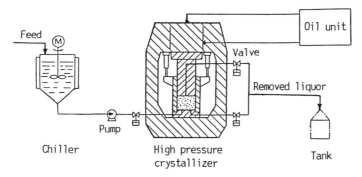

Figure 3. Typical flow diagram for an adiabatic process.

High Purity and High Recovery Rate are Attainable in One Cycle.   In an isothermal operation, experimental results showed that high purity was easily obtainable and the recovery rate was quite close to those calculated from phase diagrams in many mixtures. For examples, in the initial stage of the experiments, the 80/20 p-,m-xylene mixture was divided into 99% p-form solid and 22-25% p-form liquid without sweating. The recovery rates to the feed reached 70-75%, which were almost the same as the calculated values.
As for the sweating effect, impurities of 0.53% in a mesitylene (1-,3-,5-trimethyl benzene) were reduced to 0.02% by separations over 160 MPa. When followed by an additional separation of around 150 MPa., impurities were reduced to 0.002%. The sweating effect was thus evidently effective (4).

With an adiabatic operation, the change of state, that is, changes in temperatures, in liquid concentration, and then in solid fraction from Point A to Point C in Figure 1 can be calculated by means of thermodynamic equations together with the equation of state. The change of state during sweating and compaction(to Point D and D') is also calculated by a similar method taking the mother liquor removal into consideration (2,3). In other words, the purity,the recovery rate as well as the cumulative amount of sweating can be theoretically calculated. In an ordinary operation, the total amount of sweating will be between five and ten percent of the initial solid fraction.

Operational results of the pilot plant and p-cresol production plant showed a good agreement with the calculated values. Differences were within a few percent. The reason for this agreement is attributed to the uniformity of the state inside the vessel.

An analysis of obtained cakes and an observation of pinholes in them revealed that the sweating progressed uniformly even in a vessel with a large diameter. In addition,in high pressure crystallization, there is no need for stirring in the crystallizer. When stirring is needed, the slurry content has to be limited for about twenty percent or so. There is no such limitation in this process. Sometimes, the solid fraction could exceed 50% of the feed even in the adiabatic process. This is also due to the uniformity of the pressure.

In the adiabatic procedure, it can be said that the initial separation from a higher pressure and rather lower temperature improves the recovery rate, while from a higher pressure and higher temperature, a higher purity product is obtained. In both cases, the high pressure plays an important role in crystallization.

Short Cycle Time.   Figure 4 shows the temperature of benzene in a vessel 50 mm in diameter. Several thermocouples were arranged from the near inside wall of the vessel to the center. When pressure was applied rapidly, the temperatures rapidly rose and reached a certain value. This value was the equilibrium temperature of solid and liquid at that pressure. This fact shows that crystallization progresses uniformly and rapidly with an increase in pressure, and stops soon after the pressurization ceases. Meanwhile, the temperature begins to decrease from the outside. It was found that in many other mixture systems, crystallization progresses as rapidly as it does in benzene.

Figure 5 shows the liquid pressure change in a cycle obtained in our pilot plant. The vessel capacity was 1.5 liters and the pressure

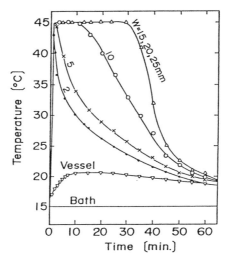

Figure 4. Temperature of benzene at points distributed in a radial direction in a vessel after pressurizing up to 156 MPa. W is the distance from the inner-surface of the vessel.

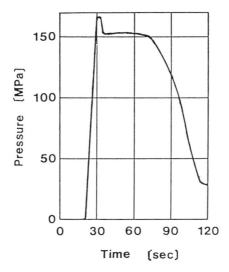

Figure 5. Pressure behavior of the liquid obtained by the pilot plant (Capacity: 1.5 liters).

could reach up to 250 MPa. In this experiment, p-xylene was separated from the p-,m-mixture of 80/20. It took 20 seconds for feeding, and 10 seconds for pressurization. After the pressurization, the mother liquor was soon separated followed by sweating and compaction. In total, it took only two minutes (2).

The p-cresol production plant completed a production run every 4.5 minutes through out the year obtaining a product of 99.5% purity from a p-,m-cresol mixture for almost the past ten years, even though its capacity was about ten times that of the pilot plant. This cycle time may be shortened even even further in the near future.

Because of the short cycle time, a large amount of production can be obtained with a vessel of relatively small capacity. For example, if an apparatus with a capacity of 20 liters runs in three-minute cycle for 7500 hours a year, 3000 tons of feed will be processed.

The short cycle time in this process results from the uniformity of pressure and the small volume change during pressurization.

Low Energy Consumption.   The total net energy for high pressure crystallization consists of energy used in the chiller E_1, for the pressurization E_2, and for the separation E_3. E_1 and E_2 were compared with the energy used in cooling crystallization. A comparison was carried out with a cresol mixture, the phase diagram of which is shown in Figure 6. Results are shown in Table I. In the table, Cooling(B) is not a real case, because the final temperature is lower than the eutectic point, 276K, and the solid fraction is too high to stir. It is cited only for comparison. In Cooling(A), the solid fraction is close to the limit for stirring.

Evidently, high pressure crystallization requires only a small amount of energy, especially in the step of pressurizing. This is not only because the decrease in volume due to pressurization is small, but also because it is compensated for by a thermal expansion due to a rise in temperature.

In this comparison, heat loses at piping and in other places and stirring energy were omitted. If they had been taken into calculation, the difference would have increased further. As for the energy for separation E_3, the work $PV_{liq.}$ should be compared with the energy of the centrifugal separator, for example, in cooling crystallization.

Simple Flow System.   In this process, the crystallization, the separation and purification are all carried out in a high pressure vessel. There is no need for respective apparatus and conveyer apparatus between them, as there is with cooling crystallization. Because high purity is obtainable in the high pressure crystallizer, washing with solvent and recrystallization are not required.

Such a simple system is also an important feature of high pressure crystallization.

Effects on Phase Diagrams

It is said that, when pressure is applied to matter, the matter undergoes the following responses,
1) decrease in volume,
2) decrease in entropy,
3) microscopic ordering.

Table I. Comparison of calculated net energies for crystallizing one mole p-cresol from the m,p-mixture of 70% p-form. Cooling (B) is an imaginary process, only shown for comparison.

	Condition	Solid fraction	Energy(kJ/mol)	
Conventional process				
Cooling (A)	298K →279.5K	0 → 0.20	36.6	
Cooling (B)	298K →272K	0 → 0.35	28.5	
High Pressure process				
Pre-cooling	298K →284K	0 → 0.06	10.9	Sum.
& Pressurizing	(0.1MPa →200MPa)	0.06→ 0.35	1.6	12.5

The last one means that the microscopic structure of the matter changes into a more structured form. Crystallization by pressure, itself, is evidently the result of this response.

The changes in the phase diagram under pressure reflect the above responses, and these changes are utilized for crystallization in many cases.

Materials Difficult to Crystallize by Conventional Means.
P-cumin aldehyde does not crystallize usually even under the temperature of the dry ice-ethanol bath. Only a glassy solid is obtained. Therefore, no melting point can be found in chemical handbooks.

Figure 7 shows the crystallization procedure of this substance. The horizontal axis shows the cumulative heat generation on an arbitrary scale. By pressurization of p-cumin aldehyde, a compression heat appeared. And when the pressure exceeded 400-500 MPa, a very small volume decrease and additional heat generation different from compression heat could be continuously seen for more than several hours. When the pressure was decreased gradually, a large amount of heat was suddenly generated accompanied by a volume decrease. This was due to crystallization. The initial change was thought to be a glass transition. Nuclei, if they were in the glassy phase, could not grow. They can grow only under a lower pressure level, while nucleation results only in or from the glassy state under a very high pressure. Even when the sample was held under the above crystallization pressure level for an entire day, crystals did not appear (*5*).

Once crystals appeared, their fraction in the system was easily controlled by varying the pressure. Thus, p-cumin aldehyde can be purified by crystallization with the use of seed crystals under pressure. The same phenomenon could be seen in m-cumin aldehyde.

As mentioned above, an excess of pressure application is useful for some substances which are hard to crystallize. Another important means is pressure swinging. This is useful not only for nucleation, but also for the acceleration of the growth rate of certain substances.

Shift in the Eutectic Concentration.   Eutectic concentration determines the theoretical limit of the recovery rate. If it shifts towards reduction of component 1, its recovery rate could increase.

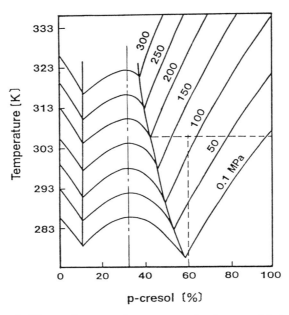

Figure 6. Phase diagram of p-, m-cresol mixture with isobars.

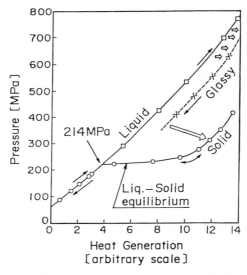

Figure 7. Cumulative heat generation of p-cumin aldehyde during liquid-glass-crystal transition.

Thermodynamics has taught us that, in a simple binary eutectic system, if the pressure dependency of the melting point of component 1 is higher than that of component 2, the eutectic concentration shifts towards component 2. That is,

if $(dT/dP)_1 > (dT/dP)_2$, then $(dX_1/dP)_e < 0$.

However, this is deduced by neglecting the second order terms, while the pressure range in this process is too wide. Therefore, this concept is not applicable here. In fact, a shift in the opposite direction to the theoretical one can be seen in many systems. It was found that a significant concentration shift rarely occurs in simple eutectic systems. We detected only one example of this significant shift, and it is shown in Figure 8(6).

Change in Solid-Solution System.   The solid-solution system is one of the most difficult systems to crystallize. A component belonging to this system is purified only by repeated recrystallization. On the left in Figure 9, the phase diagram of Br-and F-benzene system under atmospheric pressure is shown. Evidently, it belongs to the eutectic system, and, in the Br-benzene rich region, the distribution coefficient of F-benzene is 0.71. The graph on the right shows the observed results in isotherm, at 273K. Shaded and light circles show the observation results of the liqidus line by different methods. Light triangles show an analysis of solids after separation at respective pressures, and shaded ones are those after sweating. The solidus line will be near the triangles (7).

Obviously, this system is changing from a solid-solution system to a eutectic system. The distribution was estimated to be between 0.03 and 0.1. Under higher temperatures and higher pressures, the shift to a eutectic system will be more pronounced.

Even though it does not shift to a perfect eutectic system, recrystallization is rather easily carried out in the high pressure crystallization. That is, before the final stage of separation, only the pressure is decreased in order to melt, and increased again to re-crystallize, followed by a final separation.

This phenomenon is a typical example of microscopic ordering due to pressure. The shift to a eutectic system, the effect of sweating on purification, and the easy recrystallization by pressure are all very useful for application to the solid solution system.

Diminishing the Intermolecular Compound.   In the phase diagram of p-, m-cresol mixture, Figure 6, there is an inter-molecular compound of one p-, and two m-form. A eutectic point is on each side of it. Under high pressures, the eutectic point on the right shifts so as to diminish the molecular compound (8). From the feed of 60% p-form, for example, p-form cannot be separated at atmospheric pressure. While, under 200 MPa, about 30% of p-form can be theoretically recovered as crystals. Therefore, the tendency to diminish the intermolecular compound is very useful in crystallization. As crystals of intermolecular compound usually have a lower density than the component crystals, the latter is thought to be more stable under high pressure.

A similar tendency can be seen in other intermolecular compound systems, an example of which is shown in Figure 10(9). Racemic

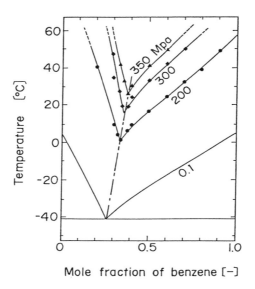

Mole fraction of benzene [−]

Figure 8. Phase diagram of benzene-cyclohexane system.

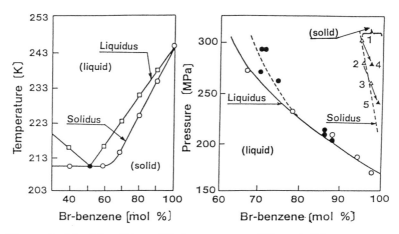

Figure 9. Liquid-solid equilibrium curves of Br-and F-benzene at
atmospheric pressure(left) and in isotherm at 273K(right).

compounds in optical isomer mixtures is also an example of the intermolecular compound. This is discussed elsewhere in this book (*10*).

Polymorphisms. In the phase diagram of oleic acid, the melting pressure of β-form is lower than that of α-form as seen in Figure 11. However, when pressure was applied, only α-form crystals appeared and increased, but never β-form crystals were obtained. As the pressure was decreased gradually, the α-form crystals melted away, and then β-form crystals appeared. As the pressure was increased again, only the β-form crystals increased, but not the α-form. α-form crystals were powder-like, β-form crystals were needle-like. These were the in-situ observations with an optical cell.

In the purification of oleic acid from a mixture containing 5% linoleic acid, the purity improved only to 96.2% in α-form, while it improved to 99.1% in β-form (*11*).

Several polymorphisms often appear in organic compounds under high pressures. The higher the pressure, the higher the density of the crystals which appear. This is one of the functions of pressure mentioned before. As the polymorphisms are easily controlled by pressure, a suitable form for separation can be selected. The above are changes in phase diagrams under high pressure available for industrial uses of crystallization. However, there are other important effects of pressure on growth rate and crystal shape or habit.

According to in-situ observations, the growth rate of many materials is tremendously high when pressure is rapidly applied(*12*). The growth rate of the dendrite crystals could exceed 50cm/sec(*13*). Also, when pressure is held constant, irregular surfaces are healed to give a nice facet construction in ten minutes, for example (*14*). It is supposed that this surprising growth behavior is caused by the number of clusters which might tremendously increase in the high pressure liquid.

Conclusion

High pressure crystallization has its own unique features, caused by properties of pressure, different from those of cooling crystallization.

In bulk production, a high purity and a high recovery rate are easily attained with a relatively small vessel in the short time of only one cycle. Changes in phase diagrams suggest an expansion to new fields of application, where conventional cooling crystallization is not always used successfully .

A basic investigation of growth behavior under pressure concerned with the physiochemical analysis of the liquid structure, and with the character of the crystals grown is needed.

Furthermore, it is important to note that there is a world-class high pressure process and plant in operation for p-cresol production, running with a short cycle time throughout the year at pressures of over 150 MPa. Since the feasibility of the process has been satisfactorily certified, it is expected that this process will find applications in various fields of the industrial separation of chemicals.

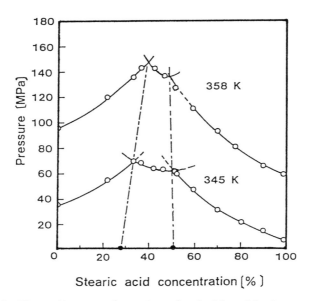

Figure 10. Phase diagram of stearic and palmitic acid mixture system in isotherm. The shaded circles are eutectic points at atmospheric pressure.

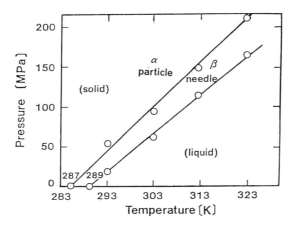

Figure 11. Phase diagram of oleic acid including 5% linoleic acid.

Literature Cited.

(*1*) Moritoki, M. In *Industrial Crystallization'84*;Jancic,S.J.,Ed.;Elsevier Sci.B.V.: Amsterdam,1984,pp.373.

(*2*) Moritoki, M.;Kitagawa,K.;Onoe,K.;Kaneko,K. In *Industrial Crystallization'84*; Jancic,S.J.,Ed.;Elsevier Sci.B.V.; Amsterdam,1984,pp. 377.

(*3*) Moritoki, M.In *High Pressure Research on Solids*;Senoo,M.; Suito,K.;Kobayashi,T.;Kubota,H.,Eds.;Current Japanese Material Research vol. 15,;Elsevier Sci.B.V.:Amsterdam,1995,pp.213.

(*4*) Moritoki, M.;*Kagaku Kougaku Ronbunshu*,1979,vol.5,pp.79.

(*5*) Moritoki, M.;Wakabayashi,M. Japanese Pat.No.1144325, 1137955,1982.

(*6*) Nagaoka.K.;Makita,T.;Nishiguchi,N.;Moritoki,M. *Int.J.Thermodynamics* 1987, Vol.8,pp.415 -424.

(*7*) Moritoki,M,;Nishiguchi,N. In *Proc. World Congress of Chem.Eng.*; JSCE:Tokyo,1986,pp.2968.

(*8*) Moritoki,M.,Fujikawa,T. In *Industrial Crystallization '84*; Jancic,S.J.,Ed.;Elsevier Sci.B.V.:Amsterdam, 1984,pp.369.

(*9*) Nishiguchi,N.;Nishida,S.;Moritoki,M. *58th Ann.Meeting of JSCE*; JSCE:Kagoshima,1993;pp.O-117.

(*10*) Nishiguchi,N.;Moritoki,M.;Shinohara,T.;Toyokura,K. cited in this book

(*11*) Nishiguchi,N.;Nishida,S.;Imanishi,N.;Moritoki,M. In *Proc.of Industrial Crystallization '93*; Rojkowski,Z.H.,Ed.: Warsaw,1993,vol.1,pp.031.

(*12*) Nishiguchi,N.;Moritoki,M.;Tanabe,H.,In *Crystal Growth 1989*; Chikawa,J.,Mullin,J.B.,Woods,J.,Eds.:North-Holland,Amsterdam, 1990,Part 2,pp.1142.

(*13*) Nishiguchi,N.;Moritoki,M. In *Proc.Int.Sym. on Preparation of Functional Materials & Industrial Crystallization'89*; Harano,Y.;Toyokura,K.,Eds.;JSCE:Osaka,1989;pp.189.

(*14*) Moritoki,M;Nishiguchi,N. In *Crystallization as a Separation Process*;Myerson,S.A.;Toyokura,K.,Eds.;ACS Symposium Series 438;ACS.:Washinton,D.C.1990,pp.220.

Chapter 14

Semibatch Precipitation of a Metallized Dye Intermediate: Reaction and Precipitation Condition Effects on Particle Size and Filterability

Kostas E. Saranteas[1,2] and Gregory D. Botsaris[2,3]

[1]Chemical Synthesis Laboratory, Polaroid Corporation, Cambridge, MA 02139
[2]Department of Chemical Engineering, Tufts University, Medford, MA 02155

An experimental study of a semi-batch reaction precipitation is presented. A mixture of copper phthalocyanine in chlorosulfonic acid and thionyl chloride is fed into a stirred reactor containing cold water, and the product precipitates in agglomerated form. A two level orthogonal array experimental design analysis shows that the agglomerate size distribution and slurry filterability are strongly dependent upon the chlorinating agent, the feed point location, and the reactor mixing configuration. Using phosphorus oxychloride rather than thionyl chloride as chlorinating agent increased significantly the precipitate particle size. A sub surface addition and the use of an axial type impeller also increased the resulting particle size. As a result of this analysis a 10-fold decrease (on average) in specific cake filtration resistance results after optimized conditions are implemented.

Precipitation processes are used routinely in the batch manufacturing of specialty chemicals. The precipitation step follows a preparative reaction step and is in turn followed by a solids filtration step. From the point of view of process optimization the reaction, precipitation, and solids separation should be considered together to ensure that no single process step dominates the cycle. This consideration will lead to optimum use of plant equipment and, therefore, to maximum throughput capacity of the plant. Precipitation as a unit operation, although very similar to crystallization, is much less understood. The key difference between the two unit operations is that

[3]Corresponding author

crystallization involves solids with some discrete solubility in the reference solvent, while precipitation involves solids having diminishing or very low solubilities in that medium. In a crystallization process product size control can be obtained through a careful control of solubility based batch temperature during the nucleation and growth stages of the crystals. In precipitation, however, product forms as a result of very rapid chemical reactions or phase changes making control of supersaturation and, therefore, nucleation, growth, and agglomeration almost impossible to achieve. Poorly defined precipitation conditions can lead to the formation of very fine or gel-like solids which are extremely difficult to separate by filtration.

The present study reflects the re-design of a manufacturing process for the production of a copper phthalocyanine dye intermediate through a three step process involving synthesis, precipitation, and filtration. Figure 1 shows the reaction scheme for the original process. The sulfonation of copper phthalocyanine with chlorosulfonic acid is followed by a chlorination step with thionyl chloride that leads to the formation of the tetra-sulfonyl chloride dye intermediate[1,2]. The product is isolated *via* a water precipitation (quenching) from the acidic solution followed by a filtration operation.

Experimental

The experimental set-up used for the precipitation studies is illustrated in Figure 2. A one liter Kontes jacketed glass reactor was used for the quench studies. The batch and jacket temperatures were controlled with a temperature control loop connected to a Camile® 2000 control system and a dual temperature Neslab® DC-25 unit. A Lightnin® model L1UO8F mixer was used to control agitation speed. The mixer shaft was coupled to a Teflon® shaft that allowed interchanging different types of mixing impellers inside the reactor.

Dimensional characterization of the crystals was obtained with the use of a Lasentec® M100C Particle Geometry Monitor. The Lasentec® M100 monitor uses an FBRM (Focused Beam Reflectance Measurement) technique to measure the rate and degree of change to the particle population and Particle Geometry (a function of the shape and dimension of the particles and agglomerates as they naturally exist in process). The Lasentec® FBRM provides both a count over a fixed time and a chord length distribution which changes with shape and dimension. Used collectively, the rate and degree of change to the Particle Geometry can be applied to the characterization and control of the crystal system. The reactor head was specially modified to allow the FBRM probe to be placed inside the reactor for *in situ* monitoring of the quench slurries.

A four blade glass rectangle, specially made to fit the reactor, was used as a baffle system when desired. Figure 3 shows the reactor mixing configuration and the dimensions of the two types of impellers tested. A Sage Instruments model 355 syringe pump was used to feed the reaction solution to the quench reactor at a constant rate. The syringe pump was coupled to a Teflon® tube which in turn was placed inside a glass cylinder for stabilization and for depth adjustment whenever sub-surface addition was desired.

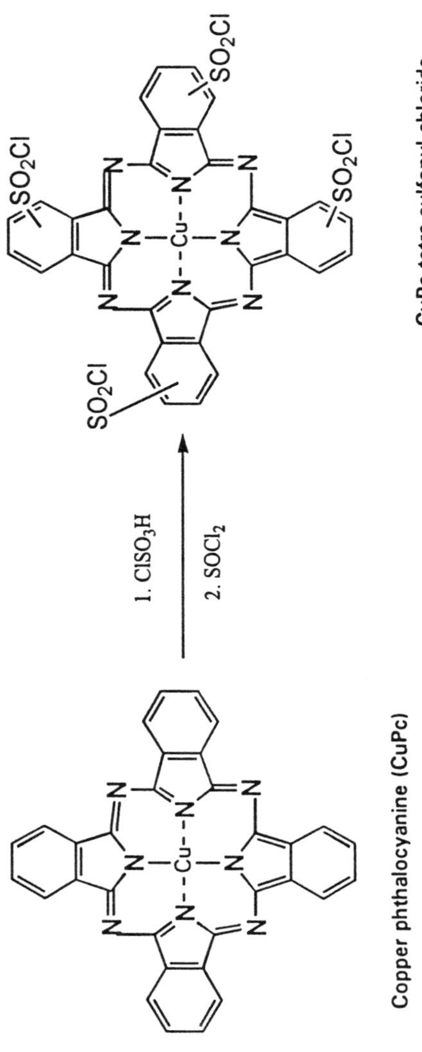

Figure 1. Reaction Scheme

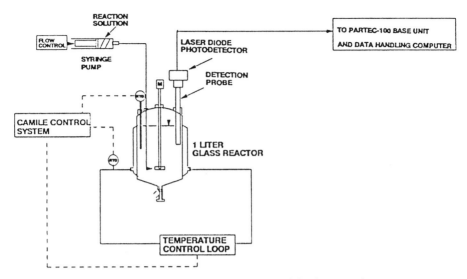

Figure 2. Experimental apparatus for precipitation experiments

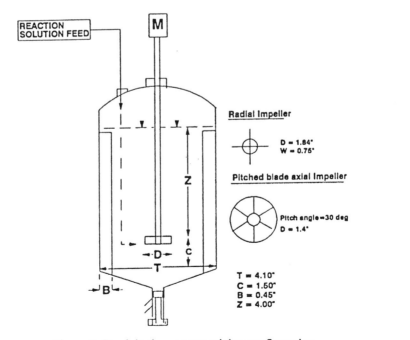

Figure 3. Precipitation reactor mixing configuration

Filterability measurement

Traditionally filterability is correlated to particle size by the Kozeny-Carman equation[3,4] expressed as a specific cake resistance, r_s, as a function of particle specific surface area S_0, solids density ρ_s and cake porosity ε:

$$r_s = \left(\frac{K_0 \cdot S_0}{\rho_s}\right) \cdot \left(\frac{(1-\varepsilon)}{\varepsilon^3}\right) \tag{1}$$

Unfortunately the use of such a relation other than illustrating first principles is extremely limited in industrial applications because both the specific area of the particles S_0 and the porosity ε are extremely difficult to characterize when dealing with agglomerated solids that are also compressible. A more useful analysis can be made to characterize the filterability of a slurry by the use of the filtration equation as defined by equation (2) below, that shows how the filtration rate is affected by the filter operating parameters (pressure drop ΔP, filtration area A, filter medium resistance R_m) and also slurry related parameters (viscosity μ, solids concentration w, specific cake resistance r_s):

$$\frac{dV}{dt} = \frac{(\Delta P)(A^2)}{\mu[r_s \cdot w \cdot V + R_m \cdot A]} \tag{2}$$

Under constant pressure, a sample slurry can be run through a filter. The filtrate volumetric throughput is determined by measuring the filtrate volume over a specific period of time that is a measured by a timer. The specific cake resistance can be estimated from a regression analysis of the integrated form of equation (2). Figure 4 illustrates a typical constant pressure filtration regression analysis. The compressibility of the solids cake can be accessed by repeating the test under different pressures. A compressibility index can be generated using a simple power function correlation as suggested by Tiller[5] of the form:

$$r_s = r_0 \cdot \Delta P^n \tag{3}$$

Based on industrial experience with filtration equipment used in the batch manufacturing plants (pressure Nutsches, basket centrifuges, filter presses) simple guidelines have been established to address the meaning of the resistance values as they relate to the ease of separation:

specific cake resistance, r_s (m/Kg)	Ease of separation
$10^{10} - 10^{11}$	easy
$10^{11} - 10^{12}$	moderate
$10^{12} - 10^{13}$	difficult
$10^{13} - 10^{14}$	very difficult

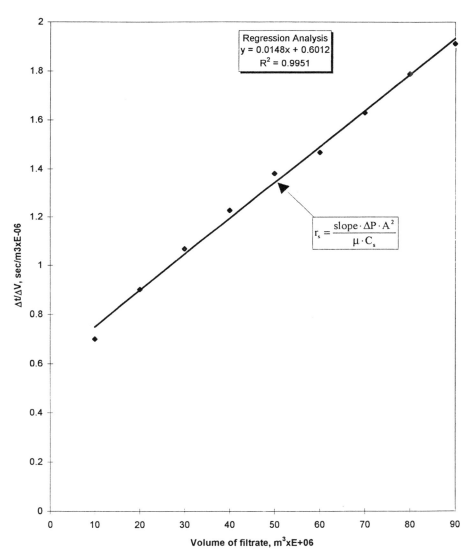

Figure 4. Typical constant pressure filtration linear regression analysis test to measure specific cake resistance

A series of filtration tests involving the standard manufacturing process generated slurry were performed to establish the ease of separation. A linear regression analysis on the constant pressure filtration data gave the following values for the filtration resistance:

Operating Pressure, psi	r_s, m/Kg
6	8.93E+11
12	2.98E+12
18	3.58E+12

Results

Quench Particle Geometry and Count   Dynamic Profiles. The particle size distribution during a semi-batch quench run, with the particle detection probe installed inside the reactor, was monitored and recorded every 35 seconds. The trends of the mean particle size, expressed either as a chord length mean or as a volume equivalent mean, resulting from such a run are illustrated in Figure 5. As can be seen, the particle size goes through a maximum at the very early stages of the quench and then asymptotically approaches a lower terminal size at the end of the quench after approximately 80 minutes. The total particle counts, as recorded by the particle size analyzer, increase monotonically with quench time at a progressively lower rate of increase until all feed is added to the reactor. Further particle size reduction and population count increase is observed as the precipitated slurry is agitated in the reactor following the addition.

Mixing Parameter Effects on Particle Size Distribution. The parameters that were considered to influence the quench slurry characteristics were the following: 1) **chlorinating agent** used in the reaction synthesis step (thionyl chloride *vs.* phosphorus oxychloride); 2) **quench feed rate** of the reaction mixture; 3) **feed point** relative to the quench water surface (above surface *vs.* below surface addition at the impeller tip); 4) **mixing impeller type** (axial *vs.* radial flow); 5) **energy dissipation** to the fluid by the mixer; and 6) **baffled** *vs.* **unbaffled** quench reactor. The agitation rate settings for the impellers were calculated using power correlations available in the literature[6] such that the low impeller power corresponded to energy dissipation of $3.4E-05$ m^5/s^3 and the high impeller power corresponded to energy dissipation of $1.1E-04$ m^5/s^3. The low and high settings for the axial impeller were 400 rpm and 600 rpm, respectively. For the radial impeller, the two settings were 190 rpm and 275 rpm.

An eight trial six factor orthogonal array experimental design was used to screen the major parameters listed above. Tables I, II describe the experimental design for the eight experiments. Two types of response parameters were determined for each experiment: a) particle size measurements from particle size analyzer (chord length mean diameter, population   balance, volume equivalent diameter); b) wet cake photomicrographic analysis.   The results obtained from the experimental design are summarized in Table III.

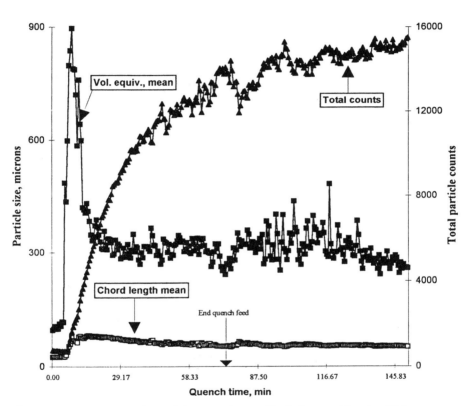

Figure 5. Dynamic particle mean size profile during semi-batch addition

Table I. Process Variable Level Assignments

Process Variable	Level 1	Level 2
X_1, Chlorinating Agent	Thionyl Chloride	Phosphorus Oxychloride
X_2, Semi-batch Feed Rate, Ml/min	4	6
X_3, Feed Point	At Impeller Tip	Above Surface
X_4, Experimental Error	-	-
X_5, Impeller Type	Radial Flow	Axial Flow
X_6, Impeller Energy Dissipation	1.1E-04	3.4E-05
X_7, Baffle Use	No	Yes

Table II. 2^7 Experimental Design for Reaction/Precipitation Analysis

Experiment	X_1	X_2	X_3	X_4	X_5	X_6	X_7
1	1	1	1	-	1	1	1
2	2	1	1	-	1	2	2
3	1	2	1	-	2	1	2
4	2	2	1	-	2	2	1
5	1	1	2	-	2	2	1
6	2	1	2	-	2	1	2
7	1	2	2	-	1	2	2
8	2	2	2	-	1	1	1

Table III. Experimental Design Final Results

Experiment	Scan count mean particle size (microns)	Volume equiv. mean particle size (microns)	Population balance (total particle counts)
1	40.5	194.4	20179
2	49.3	287.8	13649
3	48.2	237.4	17130
4	77.1	394.9	7400
5	47.1	203.1	17351
6	47.4	263.7	15208
7	40.5	150.2	20361
8	40.7	204.6	18701

A sample of an analysis of variance (ANOVA) from this experimental design is listed in Tables IV and V for two of the key response parameters. The analysis on the chord length mean diameter revealed that no factor causes variance statistically significant at the 95% level.

Table IV. Volume Equivalent Mean ANOVA Summary

Source	Df	F**	% contribution
Chlorinating agent	1	41.2	42.3
Feed point	1	26.4	27.1
Impeller type	1	21.1	21.7
Mixing power	1	5.7	5.8
Pooled error	3		3.1

**Note: $F_{c,0.05} = 10.13$

Table V. Particle Population Balance ANOVA Summary

Source	Df	F**	% contribution
Chlorinating agent	1	42.0	39.8
Feed point	1	18.4	17.4
Impeller type	1	26.0	24.7
Mixing power	1	16.2	15.3
Pooled error	3		2.8

**Note: $F_{c,0.05} = 10.13$

The results are also illustrated in Figures 6, 7 showing the average level effect of all significant factors tested. Optimum reaction and precipitation conditions can be defined as the conditions that are expected to minimize filtration resistance and, therefore, are the conditions that correspond to maximum particle size and minimum population counts. These conditions are: a) phosphorus oxychloride as a chlorinating agent; b) below surface addition of feed at the impeller surface; and c) axial type of mixing impeller. For a series of confirming runs, a mean value can be predicted for each response parameter by the equation[7] :

$$n_{opt} = \hat{n} + \sum_i (n_{A_{i,opt}} - \hat{n}) \qquad (4)$$

A prediction interval for r confirming runs around the optimum value can be defined similarly by[7] :

$$P.I. = \pm \{ F_{(1,f_e,\alpha)} \cdot V_e \cdot (\frac{1}{n_e} + \frac{1}{r}) \}^{1/2} \qquad (5)$$

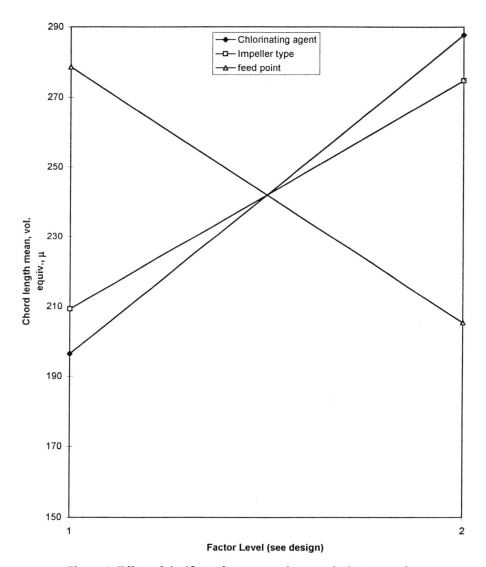

Figure 6. Effect of significant factors on volume equivalent mean size

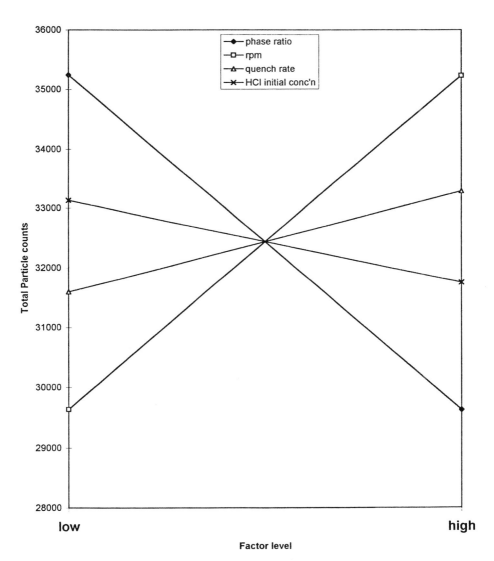

Figure 7. Effect of significant factors on particle total counts

The existence of active interactions in the experimental design can be tested with confirming runs. The prediction of response parameters as defined by equations 4 and 5 will be compared to the mean values from the r confirming runs. If they fall in the same range then most likely there are no significant active interactions.Table VI and Figure 8 show the three confirming run results. It is clear not only that the confirming runs fall within the prediction interval (no active interactions) but also the mean particle size resulting from the three runs is characterized by a much lower standard deviation than the one predicted.

Table VI. Confirmation run summary

Response parameter	95% confidence interval point estimate for 3 confirming runs	3 confirming run actual ($\mu \pm 2\sigma$)
Chord Length mean diameter	63.6±22.9	54.13±4.4
Volume equiv. mean diameter	353.5±63	331.2±15.6
Population balance (total counts)	10445±3410	11806±1120

In addition to particle size distribution obtained from the particle size analyzer, some photomicrographs were obtained to characterize further the nature of the suspensions. As can be seen in Figure 9, the precipitated dye solids are **aggregates of small spherical shaped particles**. The size of the individual particles making up the aggregates ranges from **0.01 to 0.3μm!**.

Finally, filtration tests for one of the "optimized" confirming precipitation runs were performed to compare the filtration characteristics of the slurry. Figure 10 shows a comparison of the specific cake resistance values obtained from this run and compares them to the original process filterability as defined in Figure 4. Figures 11, 12, 13 illustrate a particle size comparison of the two processes based on a chord length PSD, percent fines[*] and volume equivalent size measurement.

Discussion of results

The choice of chlorinating agent effect on agglomerate particle size distribution can be explained qualitatively rather easily if we consider the reactions taking place during the precipitation quench:

Old Process Quench Reactions

$$ClSO_3H + SOCl_2 + 2H_2O \rightarrow H_2SO_4 + 3HCl + SO_2(gas) \uparrow$$

---

[*] Percent fines is a relative frequency based process comparison for all particles with chord length mean smaller than 15 μm.

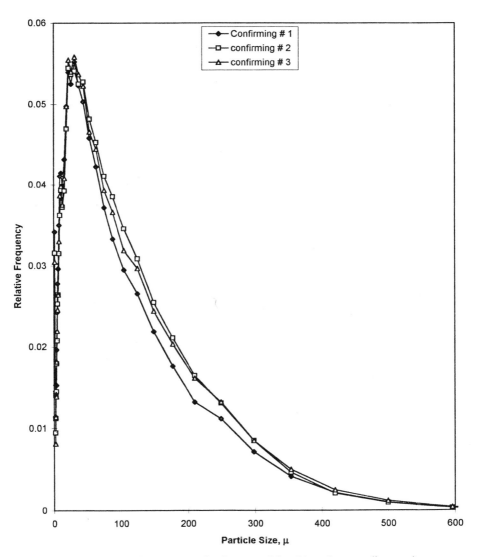

Figure 8. PSD for three confirming runs (chord length mean diameter)

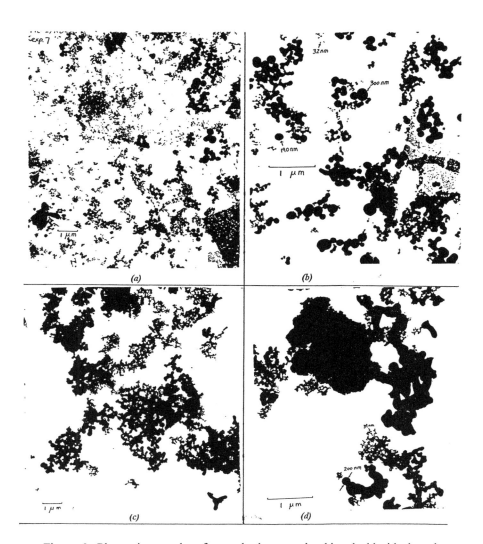

Figure 9. Photomicrographs of quench slurry. a, b: thionyl chloride based precipitation. c, d: phosphorus oxychloride based precipitation

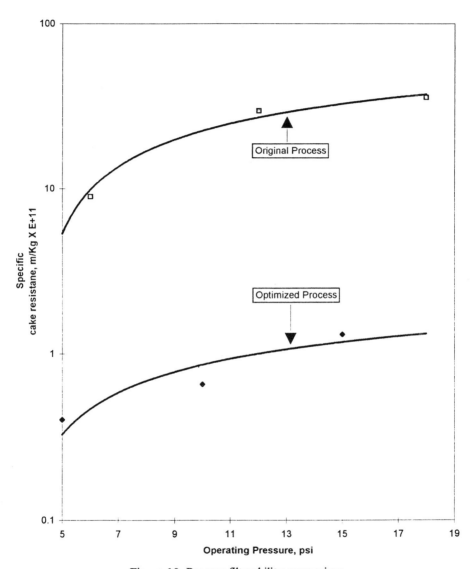

Figure 10. Process filterability comparison

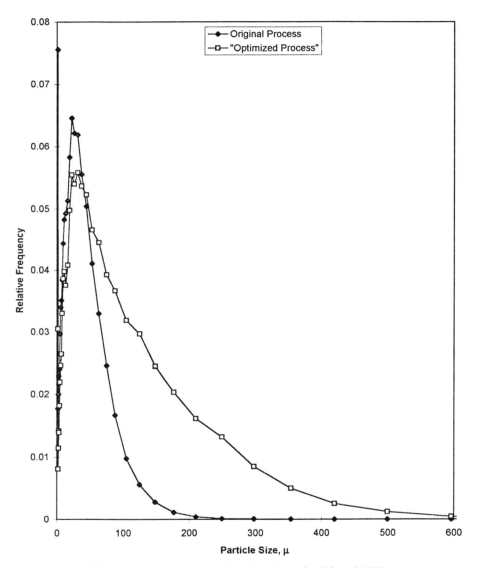

Figure 11. Process comparison based on chord length PSD

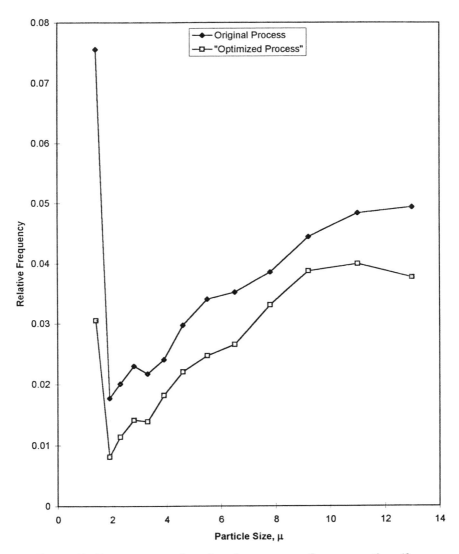

Figure 12. Process comparison based on percent fines generation (from chord length mean PSD)

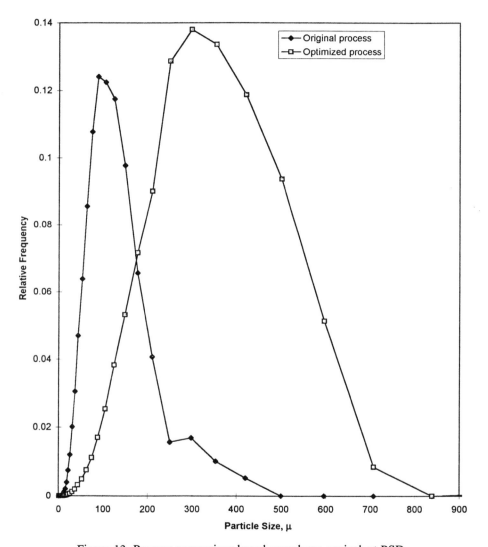

Figure 13. Process comparison based on volume equivalent PSD

Optimized Process Quench Reactions

$$ClSO_3H + POCl_3 + 4H_2O \rightarrow H_2SO_4 + H_3PO_4 + 4HCl(aq)$$

The generation of sulfur dioxide in the old process, which does not dissolve appreciably under the acidic conditions of the precipitation operation, causes a rapid gas evolution during the quench resulting in the break down of the agglomerated particles with generation of a high level of fines (as supported by volume mean and total particle counts analysis). The photomicrographs support the same conclusion showing the presence of fines and smaller agglomerate size for the thionyl chloride based precipitation process.

The mixing parameter effects are somewhat more difficult to interpret. The reaction is instantaneous in nature and the zone of reaction narrows and becomes a boundary surface between the two liquid regions. The mixing effect in the precipitation should be clear; it controls the chemical reaction rate by controlling the area of that boundary surface. The chemical reaction in turn generates product supersaturation which leads to nucleation, crystal growth and agglomeration. De-agglomeration is also expected to occur due to inter-particle collisions and fluid shear forces but is expected to be a much slower process than the other steps. Depending on which step controls (nucleation, growth, or agglomeration) we can expect different results on the final particle size and morphology characteristics.

According to B. Marcant and R. David[8] introduction of the feed in a more turbulent field should affect only the primary nucleation rate which should lead to an increase in relative particle counts and an overall decrease of particle size. However in the system we studied here we observed exactly the opposite effect: Introducing the feed in a more turbulent region actually increases the overall particle size. The most likely explanation for such an event is that agglomeration competes with nucleation as a dynamic process step. Increasing turbulence leads to further break down of the elements of the added liquid, and thus to an increased area of the boundary surface between the two liquid regions. In a fast reacting system this means higher overall reaction and precipitation rates. Increased particle concentrations result to higher agglomeration rates, since according to Smoluchowski's[9] original theory the latter has a second order dependence on particle concentration.

The power input effect on precipitation has also been discussed extensively in the literature. Nyvlt *et al.*[10]., have suggested that an increase in the power input to the system can accelerate the diffusional growth of crystals therefore increasing the crystal size. B. Marcant and R. David[8] when discussing the stirring speed, speculate that increasing speed rate enhances primary nucleation and also agglomeration rate which can have contradictory results on particle size and particle counts. The experimental results reported in the present paper suggest that increasing the stirring rate leads to increased size and also decreased total counts. This result is consistent with the previously discussed concept of accelerated agglomeration rate due to accelerated reaction rate induced by increased mixing power. However, when residence times following the precipitation step are relatively long, as is the case for scale-up of batch filtration, secondary nucleation can significantly reduce the particle size and increase the particle counts. Similar results have been reported by Tosun[11] where he observed that a combination of high stirring rate and feed addition at a more turbulent region increases particle size. The same author reported that there existed a minimum on the curve crystal size vs. impeller speed.

The fact that an axial type impeller increased the resulting particle size, as compared to the radial type impeller, can be also explained as the result of two competing actions: the mixing controlled aggregate growth by inter-particle collisions and the shear controlled aggregate break-up. Axial type impellers are characterized by low shear and high liquid flow. Thus they favor the aggregate growth over the break up. High shear, on the other hand, favors secondary nucleation and for this additional reason a radial impeller may increase the particle count at the expense of size.

Conclusions

This experimental study illustrates the complexity of the competing phenomena taking place during the semi-batch quench of an instantaneous reaction/precipitation process. Using phosphorus oxychloride rather than thionyl chloride as chlorinating agent increased significantly the precipitate particle size. A sub surface addition and the use of an axial type impeller also improved the resulting particle size. Mixing seems to affect primary nucleation, secondary nucleation, growth, and agglomeration kinetics of the resulting precipitate. At the present time, only qualitative explanations can be offered for the observed experimental results. Experimental techniques illustrated here may provide the process engineer with a method to control the solids filtration problem, but the process robustness may never be guaranteed until a fundamental understanding of all competing processes as presented above is accomplished. A follow up to this work is already on-going with particular emphasis on explaining the experimental results by mixing models (macro, meso and micro-mixing) followed by scale-up confirmation runs of the same process.

Acknowledgments

The authors would like to thank Maurizio Ciambra and Joseph Mattia of the Polaroid Chemical Synthesis Laboratory, Hei Ruey Jen of the Polaroid Research Central Analytical Laboratory, and summer intern Katherine Taylor, an undergraduate student of chemical enginnering at the Massachusetts Institute of Technology, for assistance with the experimental work.

Roman Notation

A   filtration area (m^2)
F   degrees of freedom
$F_{(1,fe,\alpha)}$ critical value of F distribution for fe degrees of freedom for the error and α risk
K_0 Kozeny constant (~5 for low porosity ranges)
n   compressibility factor

n_e defined as number of experimental runs divided by number of factors plus one

n_{opt} prediction of mean value at optimum conditions

n grant mean value of all experimental measurements

$n_{Ai,opt}$ mean value of response parameter when factor A is at optimum level

r_s specific cake resistance (m/kg)

r_0 constant related to particle size (m/kg)

R_m filter medium resistance (1/m)

S_0 specific surface of particles

t   filtration time (sec)

V   filtrate volume (m^3)

v_e Variance of error

C_s mass of dry solids per unit volume of filtrate (kg/m^3)

Greek Notation

ΔP Pressure drop across cake (N/m^2)

μ filtrate viscosity ($N\text{-}sec/m^2$)

ρ_s solids density (kg/m^3)

ε porosity of solids cake

References

1.   H., Boag A., *Photographic Products,* U.S. Patent 3,888,875 (1975).

2.   Bader, H., Boag A., *Photographic Products,* U.S. Patent 3,888,876 (1975).

3.   Kozeny, J., *Ber. Wien. Akad.,* **136a**, 271 (1927).

4.   Carman, P.C., *Trans. Inst. Chem. Eng.,* **15**, 150 (1937).

5.   Tiller, F.M., *Filtration and Separation,* **12**, 386(1975).

6.   Oldshue, J.Y., *Fluid Mixing Technology,* McGraw-Hill Publ. (1983)

7.   Wu, Y., Moore, W., *Quality Engineering* American supplier institute Inc.(1986).

8.   Mercant, B., David, R., *Experimental Evidence for the Prediction of Micromixing Effects in Precipitation,* AICHE journal, **37**,11, (1991).

9.   Smoluchowski, V., et al., *Versuch Einer Mathematischen Theorie der Koagulationskinetic Kolloider Losungen,* Z. phys. Chem. **92**, 129-168 (1917).

10.   Nyvlt, J., and M. Karel, *Crystal Agglomeration,* Cryst. Res. Technol., **20**, 173 (1985).

11.   Tosun, G., *An Experimental Study of the Effect of Mixing on Particle Size Distribution in BaSO₄ Precipitation Reaction,* Eur. Conf. on Mixing, Pravia, BHRA, 171(1988).

Chapter 15

Effects of Seeding on Start-up Operation of a Continuous Crystallizer

H. Takiyama, H. Yamauchi, and M. Matsuoka

Department of Chemical Engineering, Tokyo University of Agriculture and Technology, 24–16 Nakacho-2, Koganei, Tokyo 184, Japan

Design strategy of start-up operation was investigated to shorten the required time for start-up operation, by analyzing the changes in the crystal size distribution and highlighting the role of the size distribution of the seed crystals and the time of addition.   The main results are as follows.
(1) The crystal size distribution and time of addition of seed crystals influence the transient behavior of products crystal size distributions during start-up.
(2) The required time to attain steady state can be shortened when previous products were used as the seed crystals.

Plant structures and operations are complicated in the chemical processes because of multiple productions, energy conservation, pollution and safety.   Operation design is therefore essential to advanced operations.   For the start-up operation with great changes in operating conditions, the following items should be evaluated(*1*).
(1) Required time for start-up operation.
(2) Switching time from the off-specification product line to the product line.
    In the industrial crystallization, operation design which satisfies required product specification is also important.   However, methods of determining the seed crystals specifications i.e., size distribution and mass of seed crystals, and the time of addition have not been considered systematically.   The required time of start-up operation is known to be different by 10 residence times even in the laboratory scale depending on the seed crystals specifications(*2*).   In the present study relations between the procedure and conditions of start-up operations and the product crystal size distributions are discussed based on the analysis of product crystal size distributions.   The purpose of this study is, therefore, to obtain design strategies for start-up operation by use of a typical MSMPR (Mixed-Suspension Mixed-Product Removal) crystallizer which has been used for the theoretical analysis.   The ammonium-sulfate water system was used.

Experimental

Experimental apparatus.   Details of experimental apparatus are shown in Figure 1.   The crystallizer is a stirred vessel of 720 ml capacity water jacket.   The crystallizer is equipped with a draft tube made of acrylic resin and four baffles.   The slurry flows inside the draft tube downward.   The crystallization temperature was controlled by a programmable thermostat (EYELA NCB-3100) with a temperature sensor set inside the vessel.   The slurry was stirred with a marine type propeller made of stainless steel. Feed solution was supplied with a centrifugal magnetic pump, and its flow rate was adjusted with a needle valve.   The slurry was withdrawn by using a syringe at one minute intervals.   The syringe worked as a level controller as well as a sampler.   The slurry collected during 6 minutes was filtered every 20 minutes, and the crystal size distribution was measured by sieving after drying.   The solution concentration was measured by sampling that was carried out with a small syringe having a cotton filter every 20 minutes.

Experimental conditions.   The feed conditions were fixed as follows: The solution concentration was 44.5 wt% and the flow rate was 6.00×10^{-7} m³/s, hence, the residence time being 1.2×10^3 seconds.   The agitation speed was decided as 11.7 s⁻¹ from preliminary experiments to satisfy MSMPR conditions (*3*).   The operating temperature was 25 °C (*2*).

Start-up methods.   In this study, methods of start-up of a continuous cooling crystallizer are considered.   Although several start-up methods are possible two typical ones as shown in Figure 2 were chosen.   This description is based on the SFC (*4*) (Sequential Function Chart).   The SFC is composed of "step" and "transition". "Step", which is denoted with a box, shows the state of apparatus.   "Transition" with a bold line indicates a condition that makes a move of the "step".   The items on the side of "step" are actual operations.   The feed solution was charged to the crystallizer from the feed line.   The feed was stopped when the liquid level arrived at the set up value, i.e. the hold up was fulfilled.   Then, the solution was cooled to the operating temperature.   Seed crystals were added when the solution temperature became constant at the set up value.   After that, the continuous feeding and discharge were started.   In this study, the experimental duration was more than 10 residence times in order to ensure the attainment of steady state.   This method of start-up is defined as SU Method 1.

On the other hand, SU Method 2 is as follows.   After the solution temperature of the initial charge in the crystallizer became constant, the feeding and discharge were started.   Seed crystals were then added to the solution with a known level of supersaturation.

Four kinds of seed crystals were used and their size distributions are shown in Figure 3.   The specifications of seed crystals are as follows:
(1) No seeds.
(2) Monodisperse seed crystals at the number based average size. (△)   (The size of seed crystals being equal to the number based average size of crystals which would be obtained under the same operating condition.)
(3) Monodisperse seed crystals at the mass based average size (□), and
(4) Crystals having the product crystal size distribution which would be obtained under the same operating condition, in other words, previous products.
The amount of seed crystals was 20.0 g, equivalent to the slurry density of 27.8 kg/m³.

Judgment of steady state.   The judgment of the time needed for the response to reach the steady state is generally referred to settling time θ_s, which is necessary in order to evaluate the required time for start-up operation.   Attainment of the steady state was determined from the convergence of the crystal size distribution.   This was done by the use of the performance index (ε) which is defined as the change in the mass based crystal size distribution in every 1 residence time.

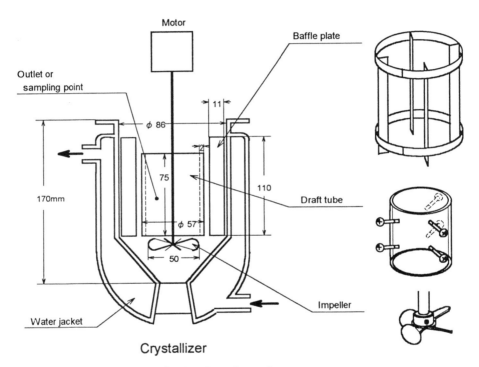

Crystallizer

Fig. 1   Experimental apparatus

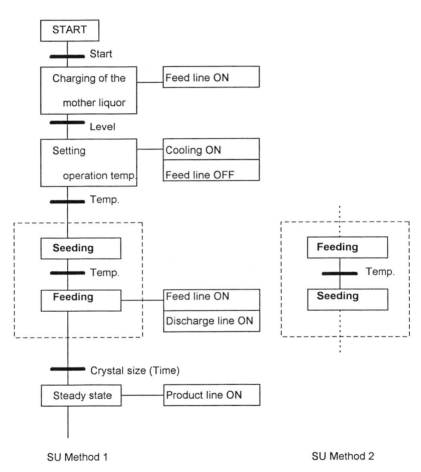

SU Method 1   SU Method 2

Fig. 2   Start-up methods

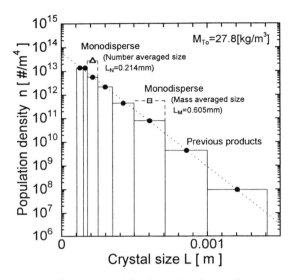

Fig. 3   Specification of seed crystals

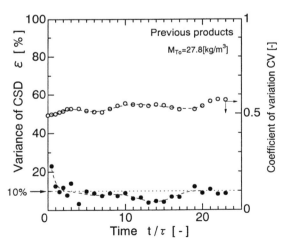

Fig. 4   Changes in ε and CV value

$$\varepsilon = \frac{\sum\limits_{j=1}^{J} w_j \cdot \left| Mp_j(t_i) \cdot dL_j - Mp_j(t_{i-1}) \cdot dL_j \right|}{\sum\limits_{j=1}^{J} Mp_j(t_i) \cdot dL_j} \cdot \frac{100}{\Delta t_i / \tau} \qquad (1)$$

$Mp_j(t_i)$ denotes the crystal mass on the j-th sieve at time t_i, and J is the total number of sieves, which is 9 in this study.   This function evaluates a change in the area of the mass based size distribution in 1 residence time. w_j is a weighting factor defined by Equation (2), and is a function of size so as to give a maximum weight to the class of the mass average size, i.e. the weighting factor of the class of the mass average size is 1.0.

$$w_j = \frac{\phi_j \cdot n(L_j) \cdot L_j^{\,3} \cdot dL_j}{\phi_M \cdot n(L_M) \cdot L_M^{\,3} \cdot dL_M} \qquad (2)$$

The above equation expresses the ratio of the mass of crystals in the j-th sieve class to the mass of crystals in the mass average size class, where, $n(L)$ is the population density function approximated with a straight line at the steady state (see Figure 3).

An experimental run with more than 20 residence times was carried out to determine the value of ε at the threshold of the attainment of steady state, and the results are shown in Figure 4.   Values of the coefficient of variation (CV) are also plotted in the diagram as ○, since the CV values are usually used to discuss the product crystal size distribution in crystallizers (5).   The CV value was calculated by Equation (3).

$$CV = \left[(m_3 \cdot m_5 / m_4^2) - 1 \right]^{1/2} \qquad (3)$$

$$m_k = \int_0^{\infty} n(L) L^k \, dL$$

The CV value of the product crystals from an MSMPR crystallizer is theoretically given as 0.5.   Since each values of ε and CV is significantly affected by errors in crystal size distribution arising from sieve analysis, the trends are approximated by smooth curves to eliminate abrupt changes and are shown as broken curves in Figure 4. The values of ε are sensitive to the changes in production quantity particularly in the beginning of the experiment, however, both the values of CV and ε converged except the beginning stage of the experiment.   From the trend of these values the threshold of the attainment of the steady state in terms of the performance index ε was decided to be 10%.   For this particular case, θ_s was therefore determined as 3 residence times. The calculated values of ε might have been affected by the relatively large crystals exfoliated from the scale on the crystallizer wall in the later stages of the experiment, and this causes fluctuations of the ε values.

Experimental Results and Discussion

No seeds.   For the case of SU Method 1 without seeds, typical changes in the crystal size distribution are shown in Figure 5.   The moment of the start of feeding was defined as time 0.   The abscissa is the dimensionless time based on the average residence time and the ordinate is the population density.   Each curve shows changes in the population density of each sieve class.   This shows that the crystal size distribution shifts to bigger size with time, and smaller crystals attained steady state faster.   Changes in ε are shown in Figure 6 for this case, and θ_s was found to be 7

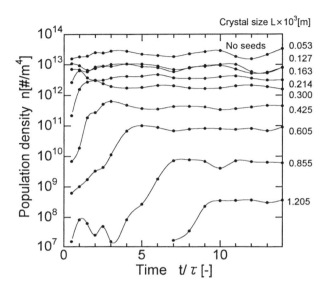

Fig. 5 Changes in population density (No Seeds)

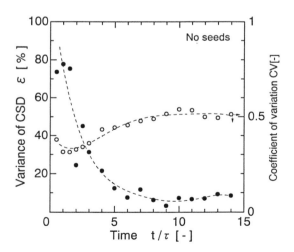

Fig. 6 Changes in ε and CV value (No Seeds)

residence times from the definition. The CV values also attained 0.5 after about 7 residence times. Changes in the average size and the production rate are shown in Figure 7. Each value was normalized by the value at the steady state. If on-line measurement of crystal size distribution was possible, the changes in the production rate for the class of the mass average size would be used to decide the switching time from off-specification product line to the product line. The average size was observed to increase continuously till about 7 residence times. The slurry density in this particular experiment was 24 kg/m^3. The total production rate also increased gradually up to 7 residence times. A decay oscillating change was observed in the production rate of the mass average size.

Monodisperse seeds. Changes in the population density when the monodisperse crystals at the number based average size were added are shown in Figure 8, where ○ shows the population density of the seed crystals. A peak arising from the seed crystals was found to shift to bigger size. It took long time for this peak to disappear. This tendency was pronounced when excess amounts of the seed crystals were added. θ_s was found to be 7 residence time after seeding. Figure 9 shows the case when the monodisperse crystals at the mass based average size were added. The movement of the peak of the seed crystals is also observed. θ_s evaluated from the ε values was 7 residence times.

Previous product seeds. Figure 10 shows the case when the seed crystals with crystal size distribution that would be obtained at the steady state were added. It is shown that the population density of each size became very stable soon after seeding. Changes in the dimensionless average size, total production rate and production rate of the class of the mass average size are shown in Figure 11. The changes of the mass average size were negligible soon after seeding. The total production rate attained stability at about 1.0 in a few residence times. The oscillated changes of the production rate at the mass average size were small compared with those of the case of no seeds (Figure 6). θ_s was 2 residence times in this case.

SU Method 2. Figure 12 illustrates the changes in the population density for the case of a different start-up method (SU Method 2). The previous product crystals were added after 1 residence time when the temperature had become constant. In this experiment, small amounts of crystals were found in the solution at about 0.5 residence time after feeding, suggesting that at the moment of seeding the solution supersaturation was large enough for nucleation. A small peak was observed in the small size region after seeding, however its movement to bigger size was not seen clearly. Changes in mass average size and the production rate are shown in Figure 13. The total production rate became stable after overshooting. The overshooting of the production rate of the mass average size, followed in about 3 residence times because of the time needed for growth of nuclei which were generated at the moment of seeding. The required time for start-up operation in this case is determined as 7 residence times as shown in Figure 14, and the changes in ε for SU Method 2 were similar to the case of no seed. It is, therefore, suggested that methods of addition of seed crystals without overshooting should be necessary in order to shorten the value of θ_s.

Conclusions

Relations between the start-up operating conditions and the specification of product crystals during start-up were experimentally examined, and the following conclusions were obtained.
(1) Differences of the size distribution, amounts of seed crystals and time of addition influence the transient changes of products crystal size distributions during start-up.
(2) The dynamic decay of oscillations of production rate of each crystal size after

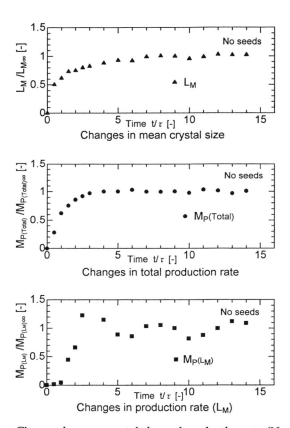

Fig. 7   Changes in mean crystal size and production rate (No Seeds)

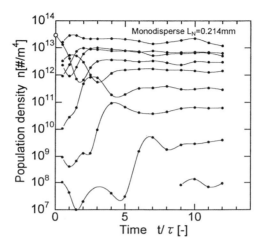

Fig. 8   Effect of seed crystals on stability of population density
          (Monodisperse seed crystals at number based average size)

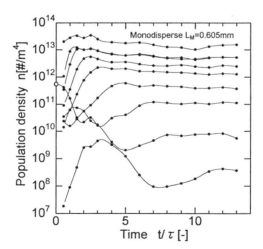

Fig. 9   Effect of seed crystals on stability of population density
          (Monodisperse seed crystals at mass based average size)

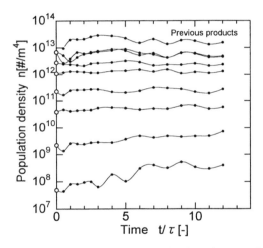

Fig. 10   Changes in population density (Previous products)

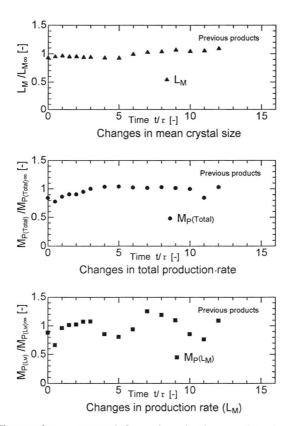

Fig. 11   Changes in mean crystal size and production rate (Previous products)

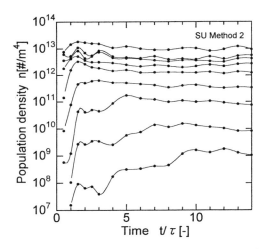

Fig. 12   Changes in population density (SU Method 2)

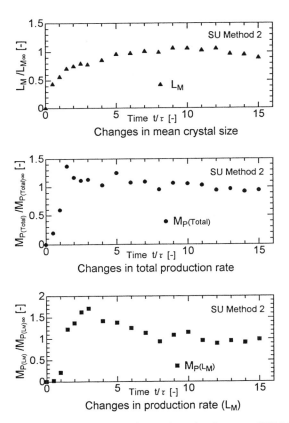

Fig. 13 Changes in mean crystal size and production rate (SU Method 2)

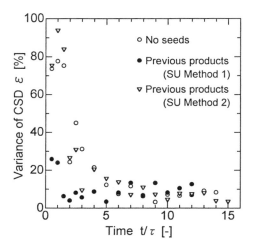

Fig. 14   Effect of start-up method on stability of crystal size distribution

start-up could be small in the case where previous products were used as the seed crystals.

(3) The time of addition did not influence the required time for start-up operation (θ_s) under the conditions of SU Method 2 and no seeds.

Nomenclature

CV	:	coefficient of variation	[-]
dL_j	:	sieve aperture width	[m]
L_M	:	mass average crystal size	[m]
$L_M/L_{M\infty}$	:	dimensionless mass average size	[-]
L_N	:	number average size	[m]
Mp_j	:	crystal mass on j-th class sieve	[kg/(m•s)]
$Mp_{(total)}/Mp_{(total)\infty}$	:	dimensionless total production rate	[-]
$Mp_{(L_M)}/Mp_{(L_M)\infty}$	:	dimensionless production rate of the class of L_M	[-]
m_k	:	k-th moment	
$n(L)$	:	function of population density	[#/m^4]
t_i	:	time of i-th sampling	[s]
w_j	:	weighting factor	[-]
ε	:	performance index	[%]
ϕ	:	volume shape factor	[-]
θ_s	:	settling time needed for the response to reach the steady state	[s]
τ	:	mean residence time	[s]

Literature Cited

1. Naka, Y.; Takiyama, H.; O'Shima, E. IFAC Workshop, London, **1992**
2. Takiyama, H.; Yamauchi, H.; Matsuoka, M. 13th Industrial Crystallization, in press
3. Matsuoka,M.; Inoue, K.; Annual Research Report of Surface & Multiphase Eng. Res. Lab, TUAT, **1984**, Vol.2
4. Baker, A.D.; Johnson, T.L.; Kerpelman, D.I.; Sutherland, H.A. Proceeding of the 1987 American Control Conference, **1987**
5. Randolph, A.D.; Larson, M.A., Theory of particulate processes 2nd.ed., Academic Press, **1985**

CRYSTALLIZATION OF PARTICULAR ORGANIC COMPOUNDS

Chapter 16

Recent Studies on Recovery of Oilseed Protein by Precipitation

Matheo Raphael[1], Sohrab Rohani[1,3], and Frank Sosulski[2]

Departments of [1]Chemical Engineering and [2]Crop Science,
University of Saskatchewan, Saskatoon, Saskatchewan S7N 5C9, Canada

Defatted oil seed meals are known to contain large amounts of nutritional proteins which can be recovered for use in food and pharmaceutical industries. Protein recovery from solution is often achieved through precipitation followed by solids-liquid separation. For efficient recovery, it is imperative to maximize the solids yield and the mean particle size, and minimize the spread of particle size distribution (PSD). In this research program, recovery of canola and sunflower protein precipitates has been investigated. Three different types of precipitators (batch, MSMPR, and tubular) and four different precipitants (HCl, carboxymethylcellulose (CMC), sodium hexametaphospate (HMP), and ammonium sulphate) have been used. Semi-Empirical models representing isoelectric precipitation of these proteins with aggregation in an MSMPRP and a dispersion model representing isoelectric precipitation of sunflower protein in a tubular precipitator are also discussed.

Recovery of Canola and Sunflower Proteins by Precipitation.   Canola and sunflower seeds are among the world's major source of edible oils. Industrially defatted meals of canola and sunflower contain about 31 to 39 mass% of protein.   These proteins are readily soluble in aqueous alkaline solutions or in aqueous solutions with low concentrations of neutral salts (salting in range) and are sparingly soluble at isoelectric pH (canola protein has two isoelectric points, at pH 3.5 and 6 and sunflower protein has an approximate isoelectric point at pH 4) or at high concentrations of salts (salting out range). This means, high amounts of protein can be extracted and

[3]Corresponding author

precipitated selectively from defatted meals, see Figure 1. Both canola and sunflower proteins contain substantial amounts of essential amino acids as shown in Table I. With further purification, these proteins can be used in food and pharmaceutical industries as substitutes of animal based proteins.

In the solution form, surfaces of protein molecules consist of charged patches (positive and negative) along with hydrophilic and hydrophobic regions. Protein molecules remain in the solution by repelling each other. The interaction of protein molecules (decrease in solubility) can be brought about by changing the solvent environment or the surface charge of the protein molecules. Decrease in solubility results in a high level of supersaturation leading to precipitation. The solvent environment can be changed by heating or by adjusting the pH of the solvent (isoelectric precipitation). Whereas, the surface charge can be changed by adding neutral salts (salting out) or organic solvents (alcohols). These remove water of hydration from the protein molecules leaving the hydrophobic regions free to combine. Details of these methods of protein recovery can be found in Bell et al. *(1)*. The choice of the precipitant depends on the final use of the protein isolate (dried precipitate), the required yield, and the type of the precipitator. For example, neutral salts are effective precipitants at very high concentrations. Thus it is easier to add solid salts in a batch precipitator than in flow-type precipitators. As shown in Table II (7), high yields of canola protein were obtained when ammonium sulphate and sodium hexametaphosphate were used as precipitants. However, the end products requires further processing to remove these salts.

For continuous operation, protein extraction using alkaline solution (pH 10 for sunflower meal and pH 12 for canola meal) followed by isoelectric precipitation using mineral acids (HCl or H_2SO_4) provide a better choice. The precipitation is fast and with proper mixing it is complete within 1 s. This is followed by aggregation of the primary particles to form large particles. The efficiency of the subsequent processes in solids protein recovery, i.e. centrifugation and spray drying, depends on the particles size and their strength to withstand the shear forces. To minimize loss of the precipitate, it is essential to have particles with a narrow particle size distribution and adequate strength.

Types of Precipitators.   Studies on precipitation of oilseeds proteins have been reported by several workers using various types of precipitators. Batch precipitators have been used by many *(2-7)*. Others *(8-11)* used mixed suspension mixed product removal (MSMPR) precipitators. Tubular precipitators have also been used *(2,12,13)*.

During protein precipitation particle size distribution evolves due to the simultaneous occurrence of the nucleation, molecular growth, aggregation, and breakup processes. Of these, aggregation is the predominant particle size enlargement process. Also, the hydrodynamics of the surrounding fluid, precipitator geometry, operating conditions, and the concentration of solids affect the particle size distribution.

In the mixed suspension mixed product removal precipitators mixing of newly formed particles with old ones results in precipitates with a wide particle size distribution. A tubular precipitator operating in the turbulent flow regime produces a narrow PSD and in the laminar flow regime a wide PSD (13). A batch precipitator

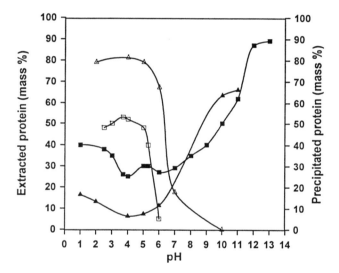

Figure 1. Extraction and precipitation yields of canola and sunflower proteins at
different pH (extraction from industrially defatted meals).
Canola: ■ Extraction yield, □ precipitation yield (7).
Sunflower: ▲ Extraction yield, △ precipitation yield.

Table I. Amino Acids Composition of Defatted Canola (Rapeseed) and Sunflower Meals *(19)*

Amino Acid	Content in defatted meal (g/16 g N)		Requirements of human adults[*]
	Canola	Sunflower	
Tryptophan	1.66	1.60	1.0
Lysine	5.86	3.12	5.4
Histidine	2.67	2.30	
Arginine	6.16	8.99	
Aspartic acid	7.34	8.78	
Threonine	4.35	3.57	4.0
Serine	4.42	4.21	
Glutamic acid	18.08	22.88	
Proline	6.13	4.98	
Glycine	5.02	5.42	
Alanine	4.34	3.92	
Methionine	2.11	1.82	
Cystine + cysteine	2.74	2.14	3.5
Valine	5.42	5.58	5.0
Isoleucine	4.34	4.67	4.0
Leucine	6.77	6.14	7.8
Tyrosine	2.59	2.27	
Phenylalanine	3.90	4.70	6.1
Ammonia	2.10	2.56	

* Source: Reference protein (FAO/WHO , 1973)

Table II. Canola Protein in solution and Mean Particle Size for Various
Precipitants in Batch Precipitator (starting protein solution at pH 12.5
with concentration 2.4 mgN/mL and volume = 30 mL)

Precipitant	Precipitant Concentration	pH	Protein in solution (mgN/mL)	Mean particle size (μm)
HCl	2 M	3.62	0.82	17.94
HMP$^+$	1.88*	3.30	0.43	31.98
CMC$^+$	0.63*	3.96	0.80	16.72
Ammonium sulphate	2.7 M	3.62	0.12	18.85

*g precipitant/g N in the feed solution (w/w)
$^+$ 2 M HCl was used to adjust the pH

results in a precipitate with a narrow PSD and with a mean particle size which depends on the type and the speed of the impeller.

Experimental Methods and Results

Experiments were performed to determine: the optimum pH for the extraction of proteins from the industrially defatted canola meal (Federated Co-op Ltd., Saskatoon, SK) and sunflower meal (Cargill Inc., West Fargo, ND), and the minimum solubility pH (isoelectric pH) for precipitation. Details of the experimental methods can be found in (7) and (13) respectively. The extracted protein solutions were then used to recover proteins by precipitation. Four different types of precipitants; aqueous HCl, HMP, CMC, and ammonium sulphate, were used for studies on the yields of canola protein. Further studies on isoelectric precipitation (using aqueous HCl as a precipitant) were carried out in three types of precipitators (Batch, MSMPR, and tubular) to study kinetic parameters and effects of precipitator type and operating conditions on PSD. The first two types of precipitators were used for canola protein and all three types were used for sunflower protein. Details of these experimental setups can be found in *(7)* for canola protein and *(13,11)* for sunflower protein.

Studies on recovery of canola protein by precipitation (near the isoelectric pH) using various types of precipitants showed that the highest yield is obtained when ammonium sulphate is used as a precipitant, but the precipitate has a small mean particle size, see Table II. The largest mean particle sizes are obtained when HMP is used as a precipitant. The lowest yield is obtained when aqueous HCl is used as a precipitant (isoelectric method). This is due to the fact that canola protein exhibits two isoelectric points (at round pH 3.5 and pH 6). Unlike canola, sunflower protein has a single isoelectric pH which leads to higher yields, see Figure 1. Figure 2 shows the scanning electron microscope (SEM) micrographs of canola and sunflower proteins obtained by isoelectric precipitations in batch precipitators. It can be seen that, both precipitates are composed of primary particles (globules of about 0.2 μm). Aggregation of these primary particles leads to the formation of larger particles. The maximum size of the aggregate depends on its strength to withstand the shear forces induced by the mixer and the surrounding fluid.

Isoelectric precipitation of sunflower protein in a batch precipitator resulted in a precipitate with small mean particle size and low coefficient of variation (CV) based on volume distribution. The population density showed a uni-modal distribution at both high and low protein feed concentrations. The mean particle size, the population density, and the CV are shown in Figures 3 to 5. The precipitation was rapid and resulted in high nucleation rates. Due to high mixing intensity, breakage rate of aggregates by shear forces and particle-particle collisions was high. This resulted in smaller particles and a narrow PSD. The steady state PSD is dependent on the hydrodynamics of the surrounding fluid. Increasing the impeller speed results in a decrease in the mean particle size due to an increase in the shear rate, see Figure 3.

For an MSMPR precipitator operating at high sunflower protein feed concentration, the mean particle size decreased as the mean residence time increased, see Figure 3, due to shear-induced breakage and particle-particle collisions. The population density showed a bi-modal PSD with a large CV at short mean residence

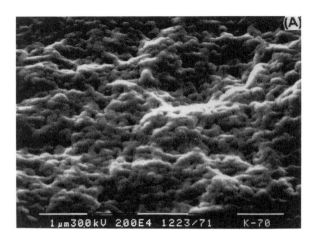

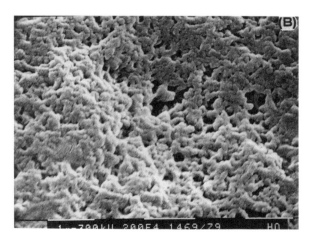

Figure 2. Scanning electron micrographs of canola and sunflower protein aggregates
(isoelectric precipitations in batch precipitators using HCl as a precipitant); (A)
canola and (B) sunflower.

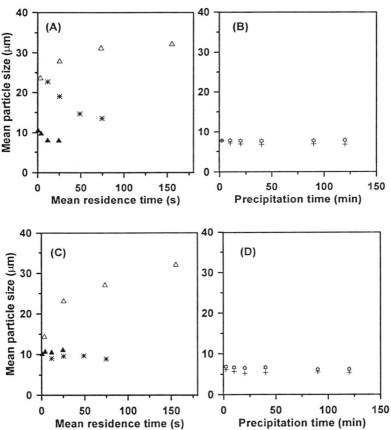

Figure 3. The effect of precipitation time on the mean particles size (sunflower protein) for the batch, MSMPR, and tubular precipitators; ⚲ = batch precipitator (600 rpm); + = batch precipitator (900); * = MSMPRP (600 rpm); Δ = tubular precipitator flow Re = 800; ▲ = tubular precipitator flow Re = 5,000. (A, B) = Protein feed concentration = 11.8 kg/m^3 (C,D) = Protein feed concentration = 2.9 kg/m^3.

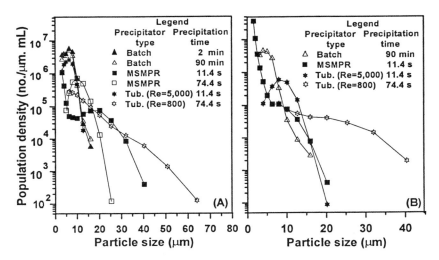

Figure 4. Population density of sunflower protein particles from different precipitators
(A) protein feed concentration = 11.8 kg/m^3 (B) = Protein feed concentration =
2.9 kg/m^3 (impeller speed for the MSMPR and batch precipitator = 600 rpm).

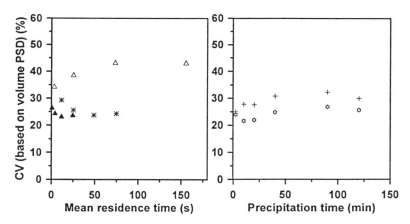

Figure 5. CV (based on volume distribution) for sunflower protein suspensions from
the batch, MSMPR, and tubular precipitators (protein feed concentration = 11.8
kg/m^3). ✿ = batch precipitator (600 rpm); + = batch precipitator (900 rpm); ∗
= MSMPRP (600 rpm); Δ =   tubular precipitator flow Re = 800; ▲ = tubular
precipitator flow Re = 5,000.

time, see Figures 4 and 5. Increasing the mean residence time transformed the bi-modal population density distribution into a uni-modal distribution with a smaller CV. The presence of newly formed particles together with aged particles may have resulted in bi-modal PSD. Earlier studies with canola protein indicate that the mean particle size increases with increases in mean residence time, see Figure 6. This is probably because a magnetic stirrer (at 260 to 425 rpm) was used in canola studies, whereas, in sunflower protein studies a three-bladed propeller (at 600 rpm) with high shear rates was utilized.

The tubular precipitator operating in the laminar flow regime resulted in larger mean particle sizes which increased with increases in mean residence time, see Figure 3. The population density distribution showed a wide spread resulting in high CV values, see Figures 3 and 5. This is due to continuous growth of smaller particles by mainly aggregation process. Low fluid-induced shear rate in the laminar flow regime favors formation of the large and stable aggregates. In the turbulent flow regime, the mean particle sizes were small and varied little with mean residence time. The population density distribution showed a uni-modal distribution with a narrow spread. In this regime, the PSD and the mean particle sizes were similar to those obtained from the batch precipitator. This suggests that fluid-induced shear determines the steady state PSD. Increasing the feed concentration under turbulent flow regime resulted in slightly smaller mean particle sizes due to increased solids concentration leading to higher aggregate breakage rate.

Figures 6 and 7 present the solids yield for canola and sunflower proteins during isoelectric precipitations, respectively. Sunflower protein yields from the flow-type precipitators increased with increases in mean residence times. This means that slower processes of particle growth by aggregation and diffusion follow an initial rapid nucleation process. About two minutes are required before the final yield is reached according to the results obtained from the tubular precipitator operating in the laminar flow regime and the batch precipitator. For canola proteins, runs in an MSMPR precipitator showed little changes in the yield with the mean residence time. This is because the mean residence times were longer (between 1.5 and 7.5 min) allowing the reaction to go to completion.

Kinetics of Canola and Sunflower Proteins Precipitations

Precipitation in an MSMPRP. Earlier studies on protein *(2,3,8,9,10,13)* have shown that, protein particles can exhibit a uni-modal or a bi-modal PSD in population density depending on the type of the precipitator and the operating conditions. The presence of large aggregates together with smaller aggregates and primary particles may result in a bi-modal PSD. In such a system, determination of kinetic parameters is difficult, because the number of primary particles changes fast due to aggregation process. *(10)* experimentally evaluated the kinetic parameters (aggregation, nucleation, and growth rates) during isoelectric precipitation of canola protein in an MSMPRP. The feed protein concentration was kept very low (0.37 to 3.1 kg/m^3) to minimize particle breakage due to particle-particle collisions. The effects of mean residence time, degree of supersaturation, and ionic strength on kinetics of nucleation, growth, and

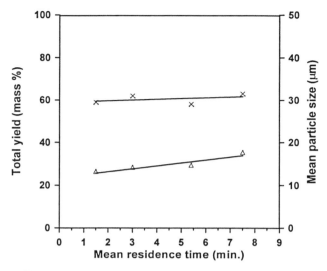

Figure 6. The effect of precipitation time on the total protein yield and mean particle
particle size for precipitation of canola protein in an MSMPRP (pH = 4.2, S =
0.0581-0.623, IS = 0.025 ml NaCl/L, impeller speed = 304 rpm). x = Yield, Δ
= Mean particle size (10).

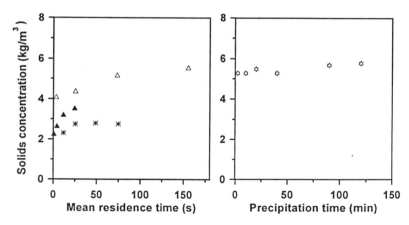

Figure 7. The effect of precipitation time on solid protein (sunflower) concentration for
the batch, MSMPR, and tubular precipitators (protein feed concentration = 11.8
kg/m^3); ✿ = batch precipitator (600 rpm); ∗ = MSMPRP (600 rpm); Δ =
tubular precipitator flow Re = 800; ▲ = tubular precipitator flow Re = 5,000.

aggregation rates were studied. The population balance for an MSMPRP operating at steady state with size independent growth is given as:

$$G\frac{dn(L)}{dL} + \frac{n(L)}{\tau} - B_{\alpha}(L) + D_{\alpha}(L) = 0, \dots n(0) = \frac{B^o}{G} \tag{1}$$

where $n(L)$ is the population density at size L, τ is the mean residence time in the MSMPRP, B^o is the nucleation rate, $B_a(L)$ and $D_a(L)$ are the birth and death rate of particles of size L due to aggregation. The birth and death rate functions were expressed using collision integral *(14)* and index of aggregation, I_{agg}. The simplified form of $B_a(L)$ and $D_a(L)$ in terms of collision frequency (β_o) are given as:

$$B_a(L) = \beta_o \frac{L^2}{2} \int_0^L \frac{n(L_1)n(\lambda)}{L_1^{2/3}} d\lambda \tag{2}$$

$$D_\alpha(L) = \beta_o n(L) m_o \tag{3}$$

$$\beta_o = \frac{1}{m_o \tau} \left[1 + \frac{4 I_{agg}}{\left(I_{agg} - 1\right)^2} \right]^{0.5} - \frac{1}{m_o \tau} \tag{4}$$

where $L_1 = (L^3 - \lambda^3)^{1/3}$ using the experimental data, the kinetic parameters were fit to the following correlations:

$$B^o = k_B \exp\left(-\frac{E_B}{RT}\right) \sigma^{\alpha_1} (IS)^{b_1} \omega^{c_1} \tau^{d_1} \tag{5}$$

$$G = k_G \exp\left(-\frac{E_G}{RT}\right) \sigma^{\alpha_2} (IS)^{b_2} \omega^{c_2} \tau^{d_2} \tag{6}$$

$$I_{agg} = k_A \exp\left(-\frac{E_A}{RT}\right) \sigma^{\alpha_3} (IS)^{b_3} \omega^{c_3} \tau^{d_3} \tag{7}$$

The results predicted from these empirical correlations agreed with experimental observations. Best agreement was obtained for the growth rate model. The discrepancies in the nucleation rate were attributed to the growth rate dispersion as the population density is extrapolated to size zero, $n(0)$.

In another study, sunflower protein with feed concentrations of (2.9 and 11.8 kg/m^3) was precipitated in an MSMPRP using aqueous HCl as a precipitant *(11)*. Due to high solids concentrations, particle-particle interactions were more frequent leading to particle aggregation and breakage. In this case, particle growth by aggregation was more pronounced than particle growth by diffusion. *(3,8,9)* studied aggregate growth of protein particles by turbulent collision mechanism and breakage by hydrodynamic shear mechanism. Based on their experimental data and previous studies on protein precipitation they developed a model for protein precipitation. The model assumes that solid protein comes out of the solution instantaneously after the precipitant is added. This results in the formation of a large number of primary particles. Growth of aggregates is by collision of primary particles with smaller aggregates. The effectiveness of collisions between primary particles and smaller aggregates is independent of their size. Collisions between larger aggregates, however, is assumed to be ineffective in forming a lasting aggregate. Larger aggregates break up to form smaller aggregates. The daughter fragments are assumed to be of equal size and their number is greater when a large aggregate breaks. Thus, in this model the growth of aggregates, G_a, is viewed as a continuous process of adding primary particles to a growing aggregate due to turbulent collisions. The primary particles are considered as growth units and larger aggregates as collectors. Shear-induced forces by the surrounding fluid or aggregate-precipitator and aggregate-aggregate collisions result in the breakage of the growing aggregates. The population balance equation written in terms of the aggregate size L is given as

$$\frac{d(G_a n(L))}{dL} + \frac{n(L)}{\tau} = B_b(L) - D_b(L) \tag{8}$$

where B_b and D_b are the birth rate and death rate of aggregates by breakage, and τ is the mean residence time. (3) proposed the following expressions for each term

$$D_b = k_d L^\beta n(L) \tag{9}$$

$$B_b = f D_b \left(f^{1/3} L \right) \tag{10}$$

$$G_a(L) = k_g L^g \tag{11}$$

Equation 10 represents the birth of particles of size L due to the death of particles of size $f^{1/3}L$ given in equation. 9. f is the number of equal-size daughter fragments. This means that the total particle volume during aggregate breakage is conserved. Substituting equations 9 to 11 in equation 8 yields

$$\frac{dn}{dL} = \frac{k_d}{k_g} L^{\beta-1} \left[f^{\beta/3 + g} n(f^{1/3}L) - n \right] - n \left[\frac{g}{L} + \frac{1}{\tau k_g L^g} \right] \tag{12}$$

with the boundary condition that the calculated total volume of aggregates equals the measured aggregate volume. All five parameters, namely, k_g, g, k_d, β, and f, were determined by optimizing a cost function given by

$$S(L_i) = \left[1 - \frac{\left(\dfrac{dn}{dL}\right)_{cal}}{\left(\dfrac{dn}{dL}\right)_{exp}} \right]^2_{L_i}$$ (13)

The best value of $S(L_i)$ is zero at all particle sizes (L_i). In the previous studies by *(9)* the values of β and f were fixed at reasonable values 2.3 and 2 for low protein concentrations (0.15 and 3 kg/m^3) and 1.5 and 3 at high protein concentrations (25 kg/m^3), respectively, and g was assumed to be 1 (linear growth rate). In our study, however, all five parameters namely, k_g, g, k_d, β, and f, were estimated. $(dn/dL)_{exp}$ was calculated from the experimental data by

$$n(L_i) = \frac{\Delta N(L_i)}{\Delta L_i}$$ (14)

where $\Delta N(L_i)$ is the number of particles/mL (as measured by the Coulter Counter) between two consecutive channels and $\Delta(L_i)$ is the channel size difference. In order to evaluate dn/dL and $n(f^{1/3}L)$ at different values of $f^{1/3}L$, it was necessary to express the experimental data in continuous form. The collocation method using piecewise fourth order cubic *spline ((15) routine DCSAKM)* was used to represent the experimental population density functions. The derivatives of this cubic spline function were calculated numerically *((15) routine DCSDER)* at selected sizes (L_i) which matched the size increments measured experimentally. The set of non-linear equations resulting from equation 9 was solved to determine the minimum sum of $S(L_i)$ using a least squares method *((15) routine DUNLSF)* or the Simplex method *((15) routine DUMPOL)*. During each iteration the value of f (number of daughter particles) was varied from 1 to 10. The set of parameters resulting in the minimum $S(L_i)$ was used to solve equation 12. The resulting ODE was solved using a sixth-order Runge-Kutta-Verner method *((15) routine DIVPRK)* to obtain the size distribution n(L). It was necessary to set the initial condition for n(L) at the smallest experimentally observed particle size. Details can be found in *(11)*.

Parameter Estimation Results. Iterations resulted in a minimum value of S when $f = 2$ in all runs. This means the number of daughter fragments, f, formed by breakage of larger aggregates was 2. This value is in agreement with that used by *(9)* Larger values of f resulted in an increase in S and lower values of predicted n(L) at the maximum size. At $f = 1$ the predicted values of n(L) were larger than the experimental values at all sizes.

The correlation between the experimental results and the calculated values for both the uni-modal and bi-model distribution is good, see Figure 8. For both, low and

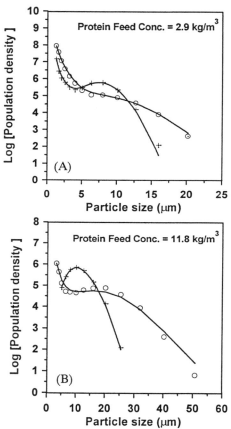

Figure 8. Population density of sunflower protein particles in an MSMPR precipitator:
Fitting of the model through the experimental data (mean residence times O =
11.4 s and + = 74.4 s); (A) Protein feed concentration = 2.9 kg/m^3 , (B) Protein
feed concentration = 11.8 kg/m^3 .

high protein feed concentration runs, the death rate constant, k_d, and the growth rate constant, k_g, decreased with increase in the mean residence time. Comparing the breakage (death) and growth rate constants, see Figure 9, shows that, at high protein feed concentrations the death process is dominant, hence resulting in smaller particles with uni-modal distribution after longer mean residence times. At low protein feed concentrations the growth process is dominant resulting in the bi-modal distribution even at longer mean residence times. This is due to the low number of collisions caused by low number of primary particles. Hence it takes longer time for the aggregates to reach the maximum stable size before they break.

The growth rate exponent, g, shows that at high solids concentrations the particle growth rate is linear with respect to particle size. At low solids concentration the growth rate exponent increases with the mean residence time showing that the growth rate is dominant.

Precipitation in a Tubular Precipitator Operating in the Dispersion Regime. Modeling of particle size distribution in an isothermal tubular precipitator involves simultaneous solution of population balance, material balance, and momentum balance equations.   For a long tubular precipitator, there exist three types of flow regimes, mixed flow regime, dispersed flow regime and plug flow regime depending on the local Peclet number.

The dispersed model after *(16)* with a modified solution technique was used to calculate the kinetic parameters from the experimental data in the dispersed zone. At steady state the population balance with no particle aggregation or breakage is given by the following differential equation

$$D_x \frac{\partial^2 n}{\partial x^2} - u_x \frac{\partial n}{\partial x} - G \frac{\partial n}{\partial L} = 0 \tag{15}$$

Nucleation rate, B^o, and growth rate, G, were expressed as functions of location and initial nucleation rate, B_o^o and initial growth rate, G_o, by:

$$B^o = B_o^o (1-z)^a \tag{16}$$

$$G = G_o (1-z)^b \tag{17}$$

where a and b are nucleation and growth parameters, respectively, and z is the dimensionless length along the reactor. Introducing dimensionless moments of population density function (f_j) and the kinetic models, the governing ordinary differential equation can be written as

$$1/Pe f_j^{(2)} - f_j^{(1)} + (0)^j (1-z)^a + j(1-z)^b f_{j-1} = 0 \tag{18}$$

with boundary conditions $f_j(0) - 1/Pe f_j^{(1)}(0) = 0$ and $f_j^{(1)}(1) = 0$. The value of $(0)^j = 1$ for $j = 0$ and 0 for $j \neq 0$, and $f^{(i)} = d^i f/dz^i$ for $i = 1$ and 2. More details can be found in *(13)*.

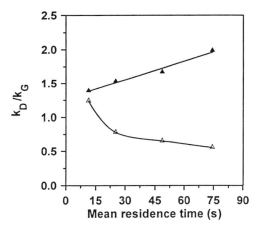

Figure 9. The ratio of breakage and growth rate constants as a function of mean residence time in an MSMPR precipitator; sunflower protein feed concentrations: $\Delta = 2.9$ kg/m^3 and $\blacktriangle = 11.8$ kg/m^3 (*11*).

Determination of Kinetic Parameters a and b. The nucleation and growth rate parameters, a and b, were obtained by minimizing the objective function, J, defined as

$$\min_{a,b} J = (CV - \hat{C}V)^2 \tag{19}$$

Where CV is the experimental value of the coefficient of variation based on weight distribution and $\hat{C}V$ is the calculated value at z = 1 ($\hat{C}V = [\{f_5(1)f_3(1)/f_4^2(1)\} - 1]^{1/2}$). The Simplex method ((*15*) *subroutine UMPOL*) was used to search for values of a and b satisfying the objective function. The tolerance limit on J was set at 10^{-12}.

Deconvolution of particle size distribution from dimensionless moments. (*17*) suggest that, four to five moments are sufficient to reproduce the original distribution using the approximate dimensionless moments as

$$f_j \approx \sum_{k=0}^{5} \xi_k^j p_k \Delta \xi_k$$

$$= \sum_{k=0}^{5} \alpha_k^j p_k \tag{20}$$

$$= [A] p_k$$

where ξ is the dimensionless particle size and p_k is the dimensionless population density function. The number of moments used in this study was six (j = 0 to 5). Therefore A is a 6 x 6 matrix (the dimensionless particle size, ξ, was divided into six nodes with the dimensionless mean size, f_4/f_3, placed in the middle of the distribution). The solution of the above system of equations gives the dimensionless population distribution function, p_k, at the selected six particle sizes. The matrix inversion method proposed by (*17*) resulted in alternating positive and negative values of p_k. Similarly, the least squares method gave positive and negative values of p_k. A method proposed by (*18*) using Laguerre polynomials with gamma-distribution weighting was employed to determine p_k. This method resulted in non-negative p_k. With known values of dimensionless population distribution functions p_k, the cumulative weight fraction distribution was calculated as

$$W(\xi) = \frac{1}{f_3(1)} \sum_{i=0}^{\zeta} r_i^3 p_i(r) \Delta r_i \tag{21}$$

where $p_i(r)$ is the dimensionless population density of particles having dimensionless size (r) smaller than ξ.

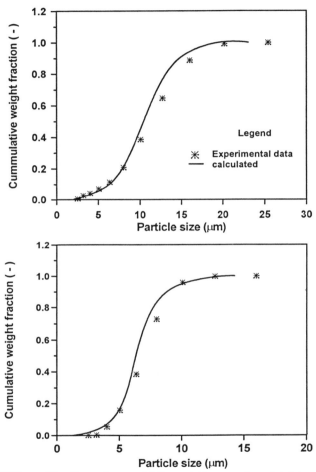

Figure 10. Fitting of the dispersion model through the experimental data: runs at flow
        Re = 5,000 (*13*).

   Parameter Estimation results. The dispersion model including the growth and nucleation models was applied to the experimental precipitation data in the dispersion range. Typical fit of the calculated cumulative particle size distribution and the experimental values for these data are presented in Figure 10. Calculated values and experimental data are in agreement within ±15%. Details can be found in *(13)*.

Conclusions

For effective recovery of solids protein from the solution it is necessary to have a precipitate with large particles and a narrow PSD. Also, to ensure maximum yield it is essential to allow particle aggregation and growth by diffusion to take place (additional time after initial nucleation). As shown experimentally in our study, it is difficult to have a precipitate with both large particles and a small CV. The tubular and the batch precipitators operating at high yields produce solids with a wide PSD and small mean particle size, respectively. The yield in an MSMPR precipitator operating at short residence times was low, the mean particle size was always small, and the PSD showed a bi-modal distribution resulting in a large CV. It appears that a batch precipitator operating initially at a high impeller speed (during nucleation) followed by a low speed (reduced shear rates) may result in a precipitate with large particles and a narrow PSD. Similarly, a tubular precipitator with a small diameter followed by an aging tank can be used to aggregate small particles to the desired PSD. Further studies are required to determine the optimum configuration.

   The semi-empirical model used for an MSMPR and the dispersion model for the tubular precipitator proved to be capable of determining the kinetic parameters (nucleation rate, aggregation rate, and growth rate) from the experimental data.

Nomenclature

a = nucleation rate parameter (-)
a_1, a_2, a_3 = supersaturation exponents in equations 5-7
b = growth rate parameter (-)
b_1, b_2, b_3 = ionic strength exponents in equations 5-7
B_a = birth rate due to aggregation (no./μm mL s)
B_b = birth rate due to breakage   (no./μm mL s)
B^o = nucleation rate (nos./cm^3s)
$B_o{}^o$ = initial nucleation rate (nos./cm^3s)
c_1, c_2, c_3 = stirrer speed exponents in equations 5-7
CV = experimental coefficient of variation (%, or fraction)
$\hat{C}V$ = calculated coefficient of variation (%, or fraction)
d_1, d_2, d_3 = mean residence time exponents in equations 5-7
D_a = death rate due to aggregation   (no./μm mL s)
D_b = death rate due to breakage   (no./μm mL s)
D_x = axial dispersion coefficient (cm^2/s)
E = activation energy (kJ/mol)

f	= model parameter, number of daughter particles formed
f_j	= jth dimensionless moment of the population density function
g	= model parameter, growth rate power
G	= particle growth rate (μm/s, or cm/s)
G_a	= aggregate growth rate (μm/s)
G_o	= initial growth rate (μm/s, or cm/s)
I_{agg}	= aggregation index
IS	=ionic strength (mol/L)
J	=objective function minimized by a and b
k_A	= constant in equation 7
k_B	= constant in equation 5
k_d	= model parameter, death rate constant
k_G	= constant in equation 6
k_g	= model parameter, growth rate constant
L	= particle or aggregate size (μm)
L_i	= internal coordinate of particle (μm)
m_j	= jth moment of population density (μm^j/L)
n	= population density (no./cm$^3\mu$m)
p	= dimensionless particle size
p_k	= dimensionless population density
Pe	= Peclet number (lu/D_x)
r	= dimensionless particle size
R	= universal gas constant (kJ/mol/K)
S	= objective function to be minimized
T	= temperature (K)
u_x	= superficial liquid velocity (cm/s)
W	= cumulative weight fraction distribution
x	= length along the tube (cm)
z	= dimensionless distance along the reactor

Greek letters

β	= model parameter, breakage power
β_o	= collision frequency (L/h)
ω	= stirrer speed (rpm)
σ	= apparent relative supersaturation
τ	= mean residence time (s)
ξ	= dimensionless particle size

Subscript

A	= aggregation
B	= nucleation
Cal	= calculated value
exp	= experimental value
G	= growth

Literature Cited

(1) Bell, D. J.; Hoare, M; Dunnill, P. *Adv. Biochem. Eng. Biotechnol.* **1983**, *vol. 26*, 1-72.

(2) Virkar P. D.; Hoare, M.; Chan, M. Y. Y.; Dunnill, P. *Biotechnol. Bioeng.*, **1982**, *vol. 24*, 871-887.

(3) Petenate, A. M.; Glatz, C. E. Biotechnol. Bioeng., **1983**, *vol. 25*, 3049-3058.

(4) Twineham, M.; Hoare, M.; Bell, D. J. *Chem. Eng. Sci.* **1984**, *vol. 39*, 509-513.

(5) Nelson, C. D.; Glatz, C. E. *Biotechnol. Bioeng.* **1985**, *vol.27*, 1434-1444.

(6) Brown, D. L.; Glatz, G. E. *Chem. Eng. Sci.* **1987,** *vol. 42*, 1831-1839.

(7) Chen, M.; Rohani, S. *Biotechnol. Bioeng.*, **1992**, *vol. 40*, 63-68.

(8) Grabenbauer, G. C.; Glatz, C. E. *Chem. Eng. Commun.*, **1981**, *vol. 12*, 203-219.

(9) Glatz, C. E.; Hoare, M.; Landa-Vertiz, J. *AIChE J.*, **1986**, *vol. 32*, 1196 -1204.

(10) Rohani, S.; Chen, M. *Can. J. Chem. Eng.* **1993**, *vol. 71*, 689-698.

(11) Raphael, M., Rohani, S. *Chem. Engng. Sci.* **1995**, *vol. 51, 4379-4384.*

(12) Chan, M. Y. Y.; Hoare, M.; Dunnill, P. *Biotechnol. Bioeng.* **1986**, *vol. 28*, 387-393.

(13) Raphael, M.; Rohani, S.; Sosulski, F. *Can. J. Chem. Engng.*, **1995**. *vol.73*, 470-483.

(14) Hounslow, M. J. *AIChE J.*, **1990,** *vol. 36*, 106-116.

(15) IMSL; *International Mathematics and Statistics Libraries Inc.;* Houston, TX,1991.

(16) Rivera, T.; Randolph, A. D. *Ind. Eng. Chem. Process Des. Dev.* **1978,***vol. 17*, 182-188.

(17) Randolph, A. D.; Larson, M. A. *Theory of Particulate Process*, 2nd ed., Academic Press, New York, N.Y., 1988.

(18) Hulburt, H. M.; Katz, S.; *Chem. Engng. Sci.*, **1964**, *vol. 19*, 555-574.

(19) Tkachuk, R.; Irvine, G. N. *Cereal Chemistry*, **1969**, *vol. 46*, 206.

Chapter 17

Secondary Crystallization of Paraffin Solidified on a Cooled Plane Surface

K. Onoe[1], T. Shibano[2], S. Uji[2], and Ken Toyokura[2]

[1]Department of Industrial Chemistry, Chiba Institute of Technology,
2-17-1 Tsudanuma, Narashino, Chiba 275, Japan
[2]Department of Applied Chemistry, Waseda University, 3-4-1 Okubo,
Shinjuku-ku, Tokyo 169, Japan

To understand and improve solids handling and granulation, a box-type crystallizer cooled by circulation of coolant through the inside was dipped into molten paraffin for a finite time. A paraffin layer was solidified on the outside plane surface of the crystallizer. Pieces of the paraffin layer were cut from the crystallizer and were stored under constant temperature, and the progress in the secondary crystallization of the solidified paraffin was observed. A correlation between the storage temperature and the change of X-ray peak strength of products was obtained for deciding operating conditions for the solidification of paraffin.

In industry, several agglomeration techniques to make solid particle have been developed. On some crystallization processes, producing crystals by solidification of a melt, the intention is to have short operation time and achieve low production cost(1,2). When operation conditions are considered, the increase of solidification rate might be favorable for improving productivity. However, there is a danger that a high solidification rate may cause inferior properties (grinding, crushing, adhering together, etc.) during a long storage. Therefore the optimum conditions should be determined by taking into account these considerations. In the case of molten paraffin solidification, there are two steps in the crystallization: the primary crystallization due to the rapid change from liquid to solid phase, and the secondary crystallization accompanying the slow change of the amorphous precipitates to crystals. These phenomena are considered to affect the properties of product crystals through their particular mechanism. In regard to the primary crystallization, the effects of the layer growth rate on the physical properties of solidified paraffin have been clarified (3). In this study, the operation conditions for the secondary crystallization of the paraffin layer produced on the cooled plane surface were varied, and a correlation between storage temperature and the change of the X-ray peak strength of the solidified paraffin was obtained.

210

Experimental

Apparatus. The schematic diagram of the experimental apparatus is shown in Figure 1. Reagent-grade solid paraffin (115P; mp 319-321K) of 0.8 kg was melted completely in the bath (1) at a temperature higher than the average temperature of the melting point by 24K, and the molten paraffin was agitated and degassed for five hours. Then the molten paraffin was cooled down and kept at melt temperature T_m (in this study 322.8K). A box-type stainless heat exchanger which has an excellent thermal conductivity was used as the crystallizer (4). The surface of the crystallizer was cooled by the circulation of an aqueous solution of polyethylene glycol through the inside of the crystallizer. The dimensions of each part of the crystallizer is shown in Figure 2. The length of the square bottom was 60mm and the depth was 15mm. Three baffles were set in the crystallizer for the efficient removal of the heat released during cooling through the square surface. The temperature changes between inlet and outlet of coolant during crystallization were measured by thermocouples.

Procedure for Primary Crystallization of Molten Paraffin. After the crystallizer was dipped in molten paraffin for a finite time, it was quickly taken out from the molten paraffin and detached from the coolant line. Then the coolant within the crystallizer was drained, and the crystal layer formed on its wall surface was kept at 298K. The primary crystallization conditions are summarized in Table 1. The coolant temperature T_c at the inlet was varied from 274.2 to 304.8K. The thickness of solidified paraffin crusted on the crystallizer surface was measured at 60 different points, and the mean value of the measurements was used for the estimation of the layer thickness L at a finite dipping time. When the observation of these items was over, the crystal layer was removed from the crystallizer and cut into about four pieces of $12 \times 15 \times 3$-5 mm in size for the secondary crystallization treatment.

Secondary Crystallization of Solidified Paraffin. The cut pieces were set in a plastic case in order to avoid direct contact between the layer surface and the case, and stored at a finite temperature within ± 0.1K by the thermostat bath. After the pieces were kept for a given time, they were taken out from the case and used for X-ray measurement to examine the structural change of the paraffin layer during the progress of the secondary crystallization. When the X-ray measurement was over, the samples were returned to their storage location and were kept at the same temperature. In all tests, the storage temperature T_k was varied from 277.2 to 305.7K.

Measurements of Physical Properties of Paraffin Layer.
    DSC Measurement. Thermal analysis of paraffin layer was performed by a differential scanning calorimeter (Seiko Inst. DSC-220). 2 mg of cut layer was set on the sample cell and heated at a constant heating rate of 5.0K/min by means of an electric furnace.

    X-Ray Analysis. X-ray diffraction patterns were obtained at three different points on both sides of the layer surface, and the mean value was used for the estimation of the peak strength I at a given storage time. Measurement conditions of the X-ray analysis are shown in Table 2. As mentioned below, the peak strength change of face (100) was chosen as a measure of the secondary crystallization rate.

Results and Discussion

Layer Growth Rate of Solid Paraffin. Typical experimental results for layer thickness of the solidified paraffin on the surface of the crystallizer are shown in Figure 3, where the time increment was set as ten-seconds intervals for the first 60 seconds of operation, twenty-seconds for the following 60 to 240 seconds and thirty-seconds for 240 to 600 seconds. The curves in the figure are data obtained for coolant temperatures

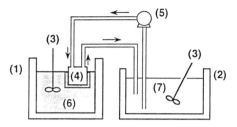

(1) Thermostat bath for melt
(2) Thermostat bath for coolant
(3) Impeller
(4) Crystallizer
(5) Circulation pump
(6) Paraffin melt
(7) Coolant (Polyethylene glycol)

Figure 1. Schematic diagram of experimental apparatus.

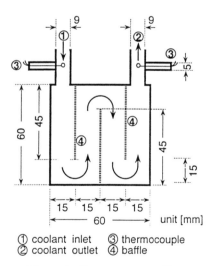

① coolant inlet   ③ thermocouple
② coolant outlet   ④ baffle

Figure 2. Box-type crystallizer.

Table 1. The Primary Crystallization Conditions

sample	: n-paraffin (115P)
melting point	: 319.2-321.2K
melting temperature T_m	: 322.8K
crystallizer	: flat box type
	$(60 \times 60 \times 15 \text{mm})$
coolant	: polyethylene glycol
coolant temperature T_c	: 274.2-304.8K
coolant flow rate	: 3.6 l/min
operation time	: 10-300 s

Table 2. X-Ray Analysis for Measurement of the Secondary Crystallization

target	: Fe
filter	: Mn
voltage	: 34 kV
current	: 14 mA
scanning speed	: 0.5-0.8 deg/min
divergent slit	: 0.5˙
scatter slit	: 0.5˙

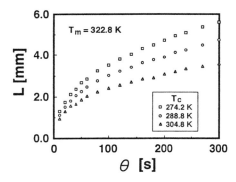

Figure 3. Examples of increase of layer thickness vs the elapsed time.

of 274.2, 288.8 and 304.8K at the inlet of the crystallizer. The increase of the coolant temperature was less than 0.2K, after the coolant passed through the crystallizer. Figure 3 clearly shows that the thickness L was inversely proportional to cooling temperature. The growth rate of the layer thickness $dL/d\theta$ was calculated from the values of $\Delta L/\Delta \theta$ of the difference of two adjacent times of operation and it was found to be 90-130μm/s for the initial ten seconds and 2-6μm/s from 240 to 600 seconds. When the crystallizer was put into the melt, the inside surface temperature of the solidified layer was the same as T_c, and supersaturation is considered to be almost (T_m-T_c) expressed as supercooling. Therefore, the average rate constant of the layer growth for the initial ten seconds is calculated for three different T_c. Their rates are 5.1μm/(s · K), 3.2μm/(s · K), and 2.7μm/(s · K) for 18.0K, 34.0K and 48.6K supercooling temperatures respectively.

DSC Patterns for Solidified and Melted Paraffin.   It has been reported that the secondary crystallite growth is being attributed to the rearrangement of carbon chains due to transition *(4)*. To measure the transition point of the produced paraffin layer, thermal analysis was performed. Figure 4a) shows the DSC pattern for heating of the paraffin layer which was produced at cooling temperature of 288.8K and kept at storage temperature of 293.2K for 20 days. Two endothermic peaks (H-①, H-②) due to dislocations in the crystals were observed in addition to the melting endothermic peak at 320.8K. Similarly, two exothermic peaks (C-①, C-②) were observed as shown in Figure 4b) when the melted paraffin in the sample cell was cooled down under the constant cooling rate of -5.0K/min by means of a gas sprayer of liquefied nitrogen.

X-Ray Diffraction Pattern.   Typical patterns of X-ray diffraction of the inside layer in the progress of secondary crystallization under the conditions of T_c=274.2K and T_k=293.2K are shown in Figure 5. The peak strength change of face (001) I_{001} and face (110) I_{110} are plotted in Figure 6. It was found that I_{001} increased from 1000 to 38000 cps (counts per second) for 37d storage, while I_{110} remained constant about 5000 cps for 59d operation. In the case of T_c=288.8K and T_k=277.2K, both I_{001} and I_{110} of the inside layer remained within the range of 3000 to 13000 cps as shown in Figure 7. In all tests by varying cooling and storage temperatures, little change was observed in the peak strength of the outside layer, indicating that the progress of secondary crystallization was negligible for the outside layer in contact with the molten paraffin and the layer growth rate was slower compared to the inside layer connected to the wall of the crystallizer.

Correlation between Cooling Temperature and Change of Peak Strength of Face (001).   The peak strength ratio I_{001}/I_{110} obtained by varying the cooling temperature is plotted against the preservation time t_k in Figure 8. On these plots, only the peak strength ratio of inside layer under the case of T_c=274.2K and T_k=293.2K was increased with passage of storage time. Figure 9 shows the correlation between layer growth rate and density of infinitely small volume ρ_p at a finite time θ, where ρ_p was calculated from the values for small increment weight ΔW and the small increment volume ΔV of a solidified layer for each interval. In these plots, the density of the inside layer at the initial step was low but those grown on the following step had near the highest density of 910 kg/m³, at almost half the growth rate at the initial step. Then, the growth rate of the layer decreased gradually with growth, and the density of the outside layer also decreased as well as the growth rate, being affected by coolant temperature.

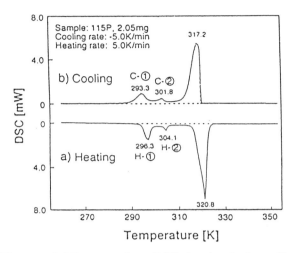

Figure 4. DSC patterns for solidified and melted paraffin.

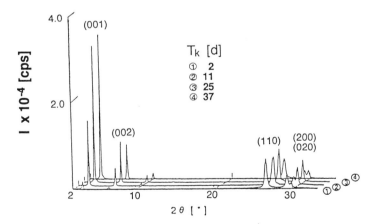

Figure 5. Typical patterns of X-ray diffraction.

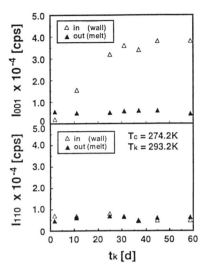

Figure 6. Comparison of X-ray peak strength change
(T_c=274.2K, T_k=293.2K).

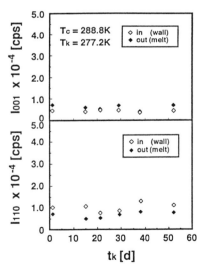

Figure 7. Comparison of X-ray peak strength change
(T_c=288.8K, T_k=277.2K).

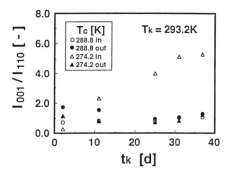

Figure 8. Dependence of T_c on I_{001}/I_{110}.

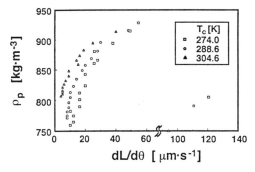

Figure 9. Effect of layer growth rate on density of infinitely small volume grown at a finite time.

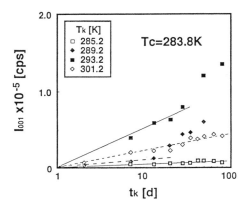

Figure 10. I_{001} vs t_k.

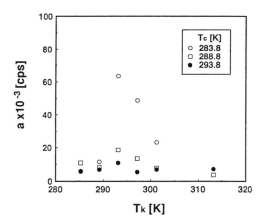

Figure 11. Relation between storage temperature and constant of X-ray peak strength change.

Correlation between Storage Temperature and Change of Peak Strength of Face (001). When I_{001} was plotted against the logarithm of storage time, Figure 10 was obtained. At storage time below 20d experimental values increased linearly against $\log(t_k)$ independently of storage temperature, and leveled up after 30d under 293.2K. It was confirmed experimentally that the change of X-ray diffraction intensity obeyed different mechanisms in the specified range of storage temperature. To examine the effect of operation conditions on the rate of structural change of solidified paraffin in the secondary crystallization stage, the correlation between storage temperature and the rate of X-ray peak strength change was calculated. The result is shown in Figure 11, where the constant a [cps] was calculated from the slope of the straight line in Figure 10. These results indicate that the constant a defined by $I_{001}/\log(t_k)$ had a highest value at storage temperature of 293.2K. Furthermore, in Figure 11, when the correlations between T_k and constant of X-ray peak strength change were plotted over the tests at coolant temperatures 283.8K and 288.8K, the maximal value of constant was obtained at T_k=293.2K, independently of the cooling temperature in this series of tests. It was found that both the temperature of cooling and that of storage turned out to affect the physical property of crystallite change. These results are useful informations about the criteria for selecting operation conditions for the production of the desired solid layer or particles.

Conclusion.

A box-type stainless crystallizer cooled inside by PEG was put in a molten paraffin and a paraffin layer was solidified on the plane surface of the crystallizer. The effects of cooling and storage temperature on secondary crystallization rate were examined, and the rate of secondary crystallization was determined from the slope of the relation between storage time and the change of X-ray peak strength of products. Furthermore, thermal analysis of the solidified paraffin was employed, and it was found that the constant of X-ray peak strength change had a maximum in a specified range of storage temperature, approximately 293K, independent of coolant temperature.

Legend of Symbols.

a	: constant defined by $I_{001}/\log(t_k)$	[cps]
dL/dθ	: layer growth rate	[mm/s]
I_{001}	:X-ray peak strength of face (001)	[cps]
I_{110}	:X-ray peak strength of face (110)	[cps]
T_c	: cooling temperature	[K]
T_k	: storage temperature	[K]
T_m	: melt temperature	[K]
t_k	: storage time	[d]

Greek Symbols;

θ	: operation time	[s]
ρ_p	: density of an infinitely small volume	[kg/m^3]

Literature Cited.

1) Ulrich, J.;   *Preprints of Waseda International Work Party for International Crystallization*, Tokyo, **1992**, pp 8-13.
2) Hunken, I.; Ulrich, J., Fishcher, O., Konig,A.   *Proceeding of 12th Industrial Crystallization*, Warsaw, **1993**, Vol.1, pp 55-60.
3) Onoe,K.; Murakoshi, K., Toyokura, K., Wurmseher, H. *Proceeding of   12th Industrial   Crystallization*, Warsaw, **1993**, Vol. 1, pp 43-48.
4) Kern,R.; Dassonville, R. *J. Crystal Growth*, **1992**, Vol. 116, pp 191-197.

Chapter 18

Crystallization of a Fatty Acid Mixture Using Spray Evaporation of Highly Volatile Solvents

K. Maeda, H. Enomoto, K. Fukui, and S. Hirota

Department of Chemical Engineering, Himeji Institute of Technology, 2167 Shosha, Himeji, Hyogo 671–22, Japan

Crystallization of an organic mixture using spray-evaporation of volatile solvents has been examined to obtain solids without inclusion of liquid. A solid mixture of lauric acid + myristic acid was dissolved in a solvent (acetone, isobutane, propane), and the solution was sprayed through a spray nozzle (100μm I.D.) and evaporated in a spray column under vacuum. The column was composed of five glass tubes, where powdery solids were collected after spray-evaporation of the solvent. Weight and composition of the solids on each glass tube were measured at various operating conditions. Most of the solids were collected on the glass tube near the nozzle and also on the bottom of the column. When the composition of the solids was analyzed, it was found that the mole fraction of lauric acid decreased from the top to the bottom of the column. The percent of solids collected on the bottom of the column was correlated well with the Weber Number, which is function of the physical properties of the solution and the operating conditions.

Attempts at crystallization of organic compounds (1) or polymeritic materials (2) using supercritical fluids have been reported in recent years. These experiments focused on the manufacture of micrometer size or sub-micrometer size particles, and on the control of particle size to obtain a uniform distribution. Highly compressed gases have been utilized as solvents to dissolve organic solids and to make supersaturation using rapid expansion, and also as anti-solvents to make the solutes in the solvents insoluble (3). Highly volatile solvents such as liquefied gases can also produce high supersaturation by evaporation. A spray-drying technique has been also used for precipitation of micro-particles (4). These solvents can dissolve not only single components but also mixtures (5), and then may be able to selectively solidify a single component from the solution. The purpose of this study was to examine the possibility of separation by crystallization using spray-evaporation of the highly volatile solvents.

Experimental Procedures

Lauric acid and myristic acid (the guaranteed reagent grade of Nacarai Tesque Inc.) were selected as the organic mixture. Acetone solvent was also a reagent of the same grade from Nacarai Tesque Inc. Isobutane and propane were liquefied gases from SUMITOMO SEIKA Inc., whose purity was more than 99 mole %. A mixture of lauric acid + myristic acid dissolved in the solvent, at a known composition, was the solution used for spray-evaporation in this investigation.

The apparatus consists of a pressurized vessel for the feed solution and a spray column, which are connected by a stainless tube, as shown in Figure 1. The vessel was set in a thermostat-bath, and the stainless tube was held to a constant temperature by a heating cable with a controller. The column consisted of five pre-divided and pre-weighed glass tubes of *0.017m I.D.* It was closed with silicon caps in both ends, and placed in vertical position. The upper cap was provided with an inlet spray nozzle of a *100 μm I.D.* and an outlet for the evaporating solvent.

The solution in the feed vessel was held to a constant temperature in a thermostat-bath. The column was evacuated to less than ten torr by a vacuum pump, and the solution was sprayed for a certain period of time to a downward direction in the column. Pressure in the column was returned to atmospheric pressure, and the five sections of the column were removed. The solids which were collected on each glass tube were weighed, and the composition of the solids was determined using gas chromatography. No fatty acids were recovered in a trap which was immersed in liquid nitrogen. This indicates a successful procedure.

Results and Discussion

To evaluate the solids crystallized by spray-evaporation, the mole weight of the solids on each glass tube was compared to the total mole weight of the solids in the column. This ratio was defined as the percent of solids collected [f], and also the composition [x] of solids on each glass tube was indicated by the mole fraction of lauric acid. The percent of solids collected and the composition of solids were analyzed at the different positions [z] along the column. Zero position ($z = 0\ m$) was defined as the top of the column, and the end position ($z = 0.11\ m$) as the bottom of it.

The solution of *0.99* mole fraction of acetone solvent, *0.10* mole fraction [x_f] of lauric acid, excluding acetone solvent, was used as the standard feed solution. The condition of *303 K* solution, with a liquid flow rate of *0.31 m/s* at the nozzle was utilized for a standard operation. The operating conditions of spray-evaporation of acetone solvent are shown in Table I.

Table I. Operating Conditions for Spray-Evaporation of Acetone

Run	t[min]	T[K]	mole fraction of acetone in feed	mole fraction of lauric acid in feed
1	10	303.2	0.99	0.1
2	10	303.2	0.98	0.1
3	10	303.2	0.999	0.1
4	10	303.2	0.99	0.1(flow rate 0.23m/s)
5	10	303.2	0.99	0.1(flow rate 0.44m/s)
6	10	303.2	0.99	0.5
7	10	303.2	0.99	0.9

Figure 2 shows the effect of liquid flow rate on the relationship between the position of the column and the composition, the percent of solids collected. A greater percent of solids was collected in the glass tube near the nozzle (0.05 m) and also on the bottom of the column (0.11m). The percent on the bottom decreased with lower liquid flow rate. The composition increased at all positions as the liquid flow rate increased. Three different mole fractions of acetone in feed are compared in Figure 3. An increase of mole fraction of acetone solvent made the percent of solids collected near the nozzle low, but made the composition of solids on the bottom high.

The effect of mole fraction of lauric acid in feed is shown in Figure 4. The percent of solids collected increased on the bottom with the following order of mole fraction of lauric acid in feed: 0.10, 0.50, 0.90. The composition of solids on the bottom, as compared with that near the nozzle, was slightly decreased in the case of 0.10 and 0.50, but it was slightly increased in the case of 0.90. The trend of the decrease in the composition along the column (purity of myristic acid in the solids increased) was considered. The isothermal solid-liquid equilibrium phase diagram for the solvent + lauric acid + myristic acid ternary system is shown in Figure 5. This process progresses rapidly as the solvent evaporates. Myristic acid is first solidifying at the point (1) and a eutectic mixture of lauric acid + myristic acid is solidifying at the point (2), secondarily. It is thought that perhaps the micro-slurry at the point (2), which was not completely evaporated, was collected on the glass tube near the nozzle, and that the powdery dry solids were not collected there, but rather fell to the bottom of the column. The powdery solids on the bottom might have also been a micro-slurry at the point (1) in the course of spray-evaporation. Most of the solids were collected near the nozzle and on the bottom of the column, as shown in Figure 2 - Figure 4. Considering spray-evaporation as a separation process, only the solids deposited on the bottom of the column may become the product in the case that lauric acid would be the impurity. It seems that the change of the composition along the column is not

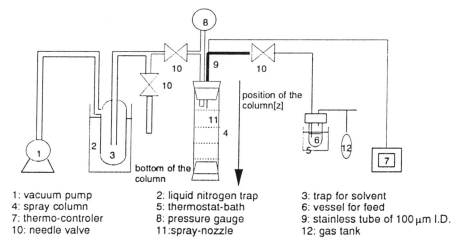

1: vacuum pump
4: spray column
7: thermo-controler
10: needle valve

2: liquid nitrogen trap
5: thermostat-bath
8: pressure gauge
11: spray-nozzle

3: trap for solvent
6: vessel for feed
9: stainless tube of 100 µm I.D.
12: gas tank

Figure 1. Experimental apparatus for spray-evaporation.

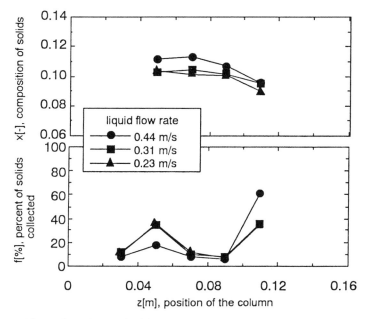

Figure 2. Effect of liquid flow rate on the relationship between position of the column and composition, percent of solids collected in spray-evaporation. (feed: mole fraction of acetone = 0.990, mole fraction of lauric acid in fatty acids mixture = 0.10)

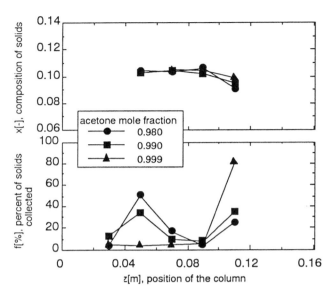

Figure 3. Effect of mole fraction of acetone in feed on the relationship between position of the column and composition, percent ofsolids collected in spray-evaporation. (feed: mole fraction of lauric acid in fattty acids mixture = 0.10)

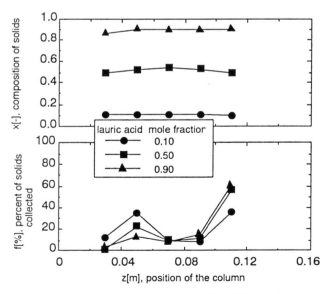

Figure 4. Effect of mole fraction of lauric acid in feed on the relationship between position of the column and composition, percent of solids collected in spray-evaporation. (feed: mole fraction of acetone ≑ 0.990)

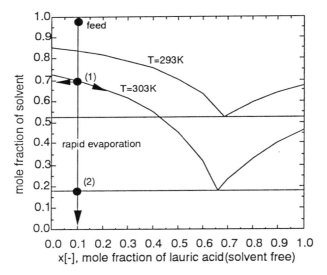

Figure 5. Isothermal solid-liquid equilibrium phase diagram for the solvent + lauric acid + myristic acid system. (ideal solution)

remarkable, because the solids collected in the column are close in composition to those collected near the nozzle and on the bottom.

To examine the role of the solvent, liquefied gases (isobutane and propane) were also used as solvents in the spray-evaporation process. The operating conditions using liquefied gases are listed in Table II.

Table II. Operating Conditions for Spray-Evaporation of Other Solvent

Run	t[min]	T[K]	mole fraction of solvent in feed	mole fraction of lauric acid in feed
isobutane				
8	0.5	298.2	0.97	0.1
9	0.5	298.2	0.98	0.1
10	0.5	298.2	0.99	0.1
propane				
11	0.5	298.2	0.97	0.1
12	0.5	298.2	0.98	0.1
13	0.5	298.2	0.99	0.1

Considering only the solids on the bottom, the relationship between separation factor and the percent of solids collected on the bottom is shown in Figure 6. The separation factor[α] was given by the following relationship:

$$\alpha = [x/(1-x)]/[x_f/(1-x_f)] \tag{1}.$$

where x_f is the mole fraction of the first crystallizing fatty acid (solvent free) in Figure 5. It was found that the fewer solids collected made the separation factor better, and that the more volatile solvent reduced the percent of solids collected on the bottom. The increase of volatility of the solvent made the separation factor close to unity. It is thought that acetone is a better solvent for spray-evaporation as a separation process, however highly volatile solvents should be required for the crystallization by spray-evaporation.

Many processes using spray techniques have been also studied using the dimensionless Weber number (6). This number is defined by surface tension [σ], liquid density [ρ], velocity [v] of the fluid, and the diameter[d] of the nozzle as

$$N_{we} = dv^2\rho/\sigma \tag{2}.$$

Figure 7 shows the effect of the Weber number on the percent of solids collected on the bottom. The percent of the solids increased with the Weber number, and the Weber number in the operations using liquefied gases were considerably larger.

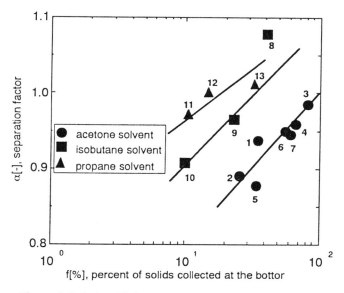

Figure 6. Relationship between the separation factor and the percent of solids collected on the bottom of the column for spray-evaporation of various solvents.

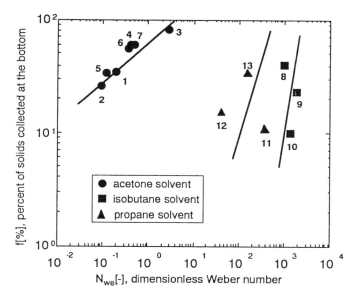

Figure 7. Effect of the Weber number on the percent of solids collected on the bottom of the column for spray-evaporation of various solvents.

Conclusions

Crystallization of fatty acids using spray-evaporation of highly volatile solvents has been examined to obtain solids without inclusion of liquid. The powdery solids were collected in the spray column, and they were usually obtained on the glass tube near the nozzle and also on the bottom of the column. The mole fraction of first crystallizing component of the solids in the column decreases from the top to the bottom of the column. The percent of the solids collected on the bottom were correlated with the Weber Number. The more volatile solvents such as propane or isobutane reduced the percent of solids and the purity of solids on the bottom. The possibility of a separation process using spray-evaporation was suggested.

Notation

x = composition of solids in the column (mole fraction of lauric acid), -
x_f = mole fraction of lauric acid in feed, -
f = percent of solids collected in the column, %
z = position of the column, m
d = diameter of the nozzle, m
v = liquid velocity at the nozzle, m/s

α = separation factor, -
ρ = density of the feed solution, kg/m^3
σ = surface tension of the feed solution, N/m

Literature Cited

1.   Chang, C. J. and A. D. Randolph, *AIChE J.*, 1989, *35*, 1876
2.   Theodore W. R., A. D. Randolph and M. Mebes and S. Yeung, *Biotechnol. Prog.*, 1993, *9*, 429
3.   Dixon D. J., K. P. Johnston and R. A. Bodmeier, *AIChE J.*, 1993, *39*, 127
4.   Bodmeier R. and H. Chen, *J. Pharm. Pharmacol.*, 1988, *40*, 754
5.   Nagahama, K., D. Hoshino, K. Maeda and M. Itoh, *Int. Chem. Eng.*, 1991, *31*, 359
6.   Grant, R. P. and S. Middleman, *AIChE J.*, 1966, *12*, 669

CRYSTALLIZATION OF PARTICULAR INORGANIC COMPOUNDS

Chapter 19

Crystallization of Sodium Chloride with Amines as Antisolvents

T. G. Zijlema, H. Oosterhof, G. J. Witkamp, and G. M. van Rosmalen

**Faculty of Chemical Technology and Materials Science,
Laboratory for Process Equipment, Delft University of Technology,
Leeghwaterstraat 44, 2628 CA, Delft, Netherlands**

The suitability of the amines diisopropylamine (DiPA) and dimethylisopropylamine (DMiPA) as antisolvents for the crystallization of sodium chloride from its aqueous solution has been demonstrated. Both amines decreased the sodium chloride solubility substantially. The presence of a two liquid phase area offered the opportunity to separate the amines from the mother liquor after crystallization by a temperature increase. In the two liquid phase area the mutual solubilities of the water and the amines were low, so the separability was good. Continuous crystallization experiments were carried out at temperatures below the liquid-liquid equilibrium line in the single liquid phase area. The product consisted of cubic agglomerated NaCl crystals with maximum primary particle sizes of 10-70 μm.

In industry highly soluble inorganic salts with a small solubility temperature dependency like NaCl, are often separated from water by evaporative crystallization. The evaporation of water in these processes is rather energy consuming. To reduce energy costs in the production of such inorganic salts, crystallization with an organic antisolvent could be an interesting alternative, because the energy intensive water evaporation step is substituted by the addition of an antisolvent. The application of antisolvent crystallization however, introduces an additional separation step, in which the antisolvent has to be recovered from the remaining mother liquor after crystallization. The antisolvent recovery must be carried out with little energy consumption in order to make the crystallization process economically feasible. With respect to this recovery diisopropylamine (DiPA) and dimethylisopropylamine (DMiPA) could serve as suitable antisolvents for the crystallization of NaCl, because they can be separated from the water phase by a temperature induced liquid-liquid phase split (1).

In Figure 1 a simplified process scheme of the antisolvent crystallization of sodium chloride is displayed. The process is divided into three steps: the crystallization, the solid-liquid separation and the antisolvent recovery or liquid-liquid separation. In the first step sodium chloride is crystallized by mixing the feed brine with an antisolvent. The crystallization is carried out at temperatures below the liquid-liquid equilibrium line in the single liquid phase area (see Figure 2). In the second step the crystals are separated from their mother liquor, e.g. by filtration or in a centrifuge. In the third and final step the antisolvent is separated from the water phase at a temperature above the liquid-liquid equilibrium line in the two liquid phase area, in which the ternary amine-water-salt system splits up into an amine and an aqueous phase. The recovered antisolvent is recycled within the process and most ideally the water phase is reused for the dissolution of crude sodium chloride. In this paper the crystallization and the liquid-liquid separation steps will be treated.

The goal of this study is to investigate whether DiPA and DMiPA are suitable antisolvents for the crystallization of sodium chloride. Key issues are the phase behaviour of the ternary amine-water-salt system (liquid-liquid and solid-liquid equilibria) and the size, shape and purity of the NaCl product formed by the antisolvent crystallization with amines.

To be able to select a crystallization temperature in the single liquid phase region at a given amine fraction, the liquid-liquid equilibrium lines of the amine-water-$NaCl_{sat}$ systems were determined. To establish to what extent an antisolvent reduces the sodium chloride solubility and to calculate the maximum obtainable magma densities during crystallization, sodium chloride solubilities in the amine-water mixtures were measured. Finally continuous crystallization experiments were carried out and the feasibility of an antisolvent recovery by a temperature increase was investigated.

Experimental

Liquid-liquid Equilibrium Lines of the Amine-Water Systems Saturated with NaCl. The liquid-liquid equilibria experiments were carried out in a jacketed 125 ml glass flask. A magnetic stirrer in combination with a stir bar provided agitation and a Lauda RK-8-KP thermostat was used to control temperature. The temperature of the vessel content was measured with an ASL precision thermometer (0.01 °C accuracy, 0.002 °C repeatability) and a small hole in the vessel cover ensured atmospheric pressure. Figure 3 shows a schematic drawing of the set-up.

The experiments were carried out by filling the vessel with known amounts of antisolvent and water, and subsequently by saturating the mixtures with excess sodium chloride at room temperature. The systems were saturated with sodium chloride during the entire experiments. Points on the liquid-liquid equilibrium line were determined in two different ways: either by visual observation of the mixing/demixing temperatures at a given overall amine to water ratio or by determining the compositions of both the organic and the inorganic phases at a fixed temperature in the two liquid phase area. The amine concentrations in the liquid samples were determined with a Chrompack

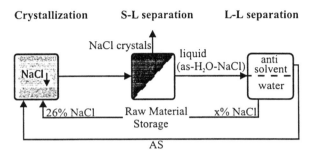

Figure 1.   Simplified process scheme of the antisolvent crystallization of NaCl.

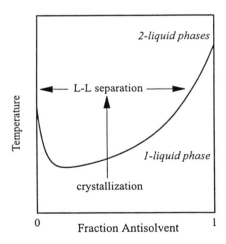

Figure 2.   Crystallization and antisolvent recovery conditions.

CP 9002 gas chromatograph equipped with a CP-Sil-5 CB column, the sodium chloride concentrations were determined gravimetrically and the water concentrations were calculated from the determined amine and sodium chloride concentrations.

When the compositions of both the organic phase and the water phase change strongly with temperature (the middle part of the phase line), points on the liquid-liquid phase line are most reliably determined by using the first method (accuracy 0.05 °C). However, when the mixing/demixing temperature is a very strong function of the amine to water ratio in the mixture (the sides of the phase line), the second method is favoured for maximum reliability.

Sodium Chloride Solubilities in Amine-Water Mixtures. The sodium chloride solubilities were determined as a function of the amine concentration at the crystallization temperature in the single liquid phase region. The experiments were carried out in the equilibrium vessel as described in the previous section. The vessel was filled with a saturated sodium chloride solution. At the crystallization temperature a known amount of amine was added, resulting in the crystallization of sodium chloride. After a period of at least 60 minutes the stirring was stopped and liquid samples of the mixture were taken. Conductivity measurements showed that this period of time was sufficient to reach chemical equilibrium. The salt concentrations in the liquid samples were determined gravimetrically.

Continuous Crystallization Experiments. Continuous crystallization experiments were carried out at temperatures in the single liquid phase region at several brine to amine feed ratios. The residence time was 30 minutes and each experiment lasted at least 8 residence times. The experiments were conducted in a continuous 1 liter crystallizer (Figure 4). The brine and the antisolvent were stored in two 10 l vessels positioned on two separate balances. From these tanks both antisolvent and brine were transported towards the crystallizer by two membrane pumps. To ensure constant feed mass flowrates the pump frequencies were controlled by dosage controllers. The feedstreams were continuously introduced into the thermostrated baffled crystallizer (4 baffles) directly above a pitched blade stirrer (N = 750 rpm). With 30 minutes intervals solid samples were taken from the crystallizer outlet, which were washed with ethanol after filtration. The amine uptake of the sodium chloride product was determined with TOC (total organic carbon) and GC, and SEM pictures (scanning electron microscopy) of the crystals were taken. The remaining part of the crystallizer outlet stream was discarded into a waste vessel. During the experiments the crystallizer contents were observed continuously with a CCD-camera, which was mounted on an Optem Zoom 70 microscope. The microscope was positioned in front of a window in the crystallizer, which was lit by a lamp from the inside of the crystallizer. The experimental set-up has been fully automated (Intellution FIX/MMI) and can be operated at temperatures of -10 to 200 °C and at pressures up to 20 bars. To prevent corrosion hastelloy-C has been used as the main construction material for the crystallizer and most of the piping. The antisolvent recovery step has not been integrated in this experimental set-up, but was investigated separately as will be described in the next section.

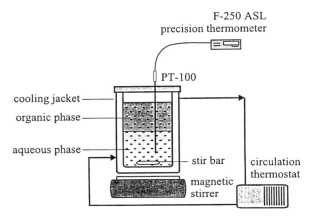

Figure 3. Experimental set-up used for determining the liquid-liquid equilibria.

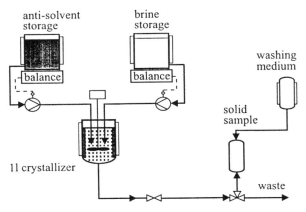

Figure 4. Schematic overview of the experimental set-up used for the continuous crystallization experiments.

Amine Recovery by a Temperature Induced Liquid-Liquid Phase Split. The antisolvent recovery experiments were conducted in the experimental set-up as described in the section 'Liquid-liquid Equilibrium Lines of the Amine-Water Systems Saturated with NaCl', and the same analytical techniques were used for sample composition determination. The experiments were carried out by filling the equilibrium vessel with amine-water mixtures of known compositions. The amine water mixtures were saturated with sodium chloride at the crystallization temperature, so that the mixture compositions equaled the mother liquor compositions in an antisolvent crystallization experiment. Subsequently at several temperatures in the two liquid phase region, samples of both phases were taken in order to determine to what extent the amine separated from the aqueous phase.

Results and Discussion

Liquid-liquid Equilibrium Lines of the Amine-Water Systems Saturated with NaCl. The liquid-liquid equilibrium lines of the binary DiPA-H_2O and DMiPA-H_2O systems are reported in the literature (2) (Figure 5). The lower critical solution temperatures (LCST) of the DiPA-H_2O and the DMiPA-H_2O systems were estimated to be 27 °C and 65 °C respectively. Below these temperatures there is one single liquid phase present regardless of the concentration of the amine.

The determined two-phase envelopes of the DiPA-H_2O-NaCl and the DMiPA-H_2O-NaCl systems (saturated with sodium chloride) are also displayed in Figure 5. The LCST's of the DiPA-H_2O-NaCl$_{sat}$ and the DMiPA-H_2O-NaCl$_{sat}$ systems were estimated to be -11.7 °C and 6.6 °C. The LCST of the DiPA-H_2O-NaCl system deviates from the one presented by Weingärtner et al. who estimated the LCST at -7.5 °C. The LCST's of the amine-water systems as well as the mutual solubilities of the water and the amines in the two liquid phase area decreased substantially as a result of the presence of sodium chloride. The dissolved sodium chloride decreases the miscibility of the amine-water mixtures, due to the fact that it prefers to be surrounded by water molecules and not by amine molecules. At lower temperatures however, the amines are sufficiently hydrophilic to be fully miscible with water saturated with sodium chloride.

With the amine-H_2O-NaCl$_{sat}$ liquid-liquid equilibrium lines the conditions were selected for the continuous crystallization experiments in the single liquid phase area. The experiments with DMiPA were carried out at X_{DMiPA} = 0.1, 0.3, 0.6 and 0.9 at a temperature of 5 °C. With DiPA sodium chloride was crystallized at X_{DiPA} = 0.9 and T = 1 °C (all the amine concentrations are expressed as salt free weight fractions).

Sodium Chloride Solubilities in Amine-Water Mixtures. To determine to what extent the amines reduce the sodium chloride solubility and to be able to calculate the maximum obtainable magma densities during crystallization, sodium chloride solubilities in the amine-water mixtures were measured. The sodium chloride solubilities were determined at the crystallization conditions in the single liquid phase area. The solubility diagram of sodium chloride in DMiPA-H_2O at T = 5 °C is displayed in Figure 6. At amine fractions above 0.8 the sodium chloride solubility approaches zero for both DiPA and DMiPA.

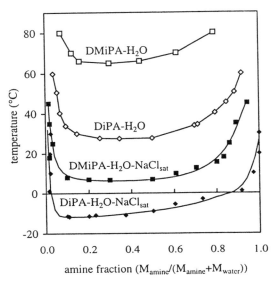

Figure 5. The two phase envelopes of the binary $DiPA-H_2O$ and $DMiPA-H_2O$ systems and of the ternary $DiPA-H_2O-NaCl_{sat}$ and $DMiPA-H_2O-NaCl_{sat}$ systems. Below the phase lines there is one single liquid phase and above there are two liquid phases present.

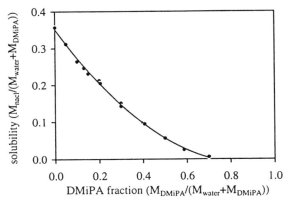

Figure 6. The sodium chloride solubility as a function of the DMiPA concentration at 5 °C.

With the experimentally determined solubility data the maximum obtainable magma densities ($M_{T,max}$) were calculated for the crystallization of sodium chloride from its saturated solution with DMiPA as antisolvent. $M_{T,max}$ is defined as the total mass of solids divided by the mass of the suspension times one hundred, if the supersaturation in the crystallizer is zero. In Figure 7 $M_{T,max}$ is displayed as a function of the amine to brine feed ratio. It can be seen that the maximum obtainable magma densities are generally low, which will probably give rise to high supersaturations during crystallization.

Continuous Crystallization Experiments. Continuous crystallization experiments were conducted in the single liquid phase area with DiPA and DMiPA as antisolvents. The size, shape and purity of the sodium chloride crystals formed in these experiments were studied. To investigate the influence of the antisolvent concentration on these product characteristics, experiments were performed at several amine to brine feed ratios. All crystallization experiments were carried out in duplicate and the reproducibility was good.

With DMiPA continuous crystallization experiments were conducted at four different amine to brine (26 wt% NaCl) feed ratios at a temperature of 5 °C. The results of these experiments are listed in Table I.

Table I. Results of the Continuous Crystallization Experiments with DMiPA

X_{DMiPA}	DMiPA Uptake Crystals	Maximum Primary Particle Size	Maximum Agglomerate Size	Shape Crystal Product
(-)[a]	(ppm)	(μm)	(μm)	
0.1	240	50	150	slightly agglomerated
0.3	40	70	200	agglomerated
0.6	220	60	150	agglomerated
0.9	630	10	100	strongly agglomerated

[a]The DMiPA fractions are calculated on a salt free basis.

In each experiment cubic agglomerated NaCl crystals were formed. The degree of agglomeration increased with the amine fraction at which the experiments were conducted. At an antisolvent fraction of 0.9 the crystal product was strongly agglomerated and consisted of small primary particles with maximum sizes of up to 10 μm (see Figure 8). The high DMiPA uptake in the large agglomerates produced at this antisolvent fraction, could well be the consequence of trapped amine rich mother liquor volumes in the voids between the primary particles in the agglomerates. At the other three antisolvent fractions the agglomerates consisted of larger primary particles with maximum sizes of about 60 μm.

Some of the sodium chloride crystals had pyramidally shaped depressions in their crystal faces. These so called hopper crystals were most frequently encountered in the experiments with $X_{DMiPA} = 0.1$. Other crystal faces contained holes, which seemed to be partly overgrown (see Figure 9). Hopper crystals are generally formed in situations where the volume diffusion of growth units to the crystal surface is rate limiting for

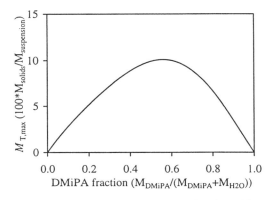

Figure 7. The maximum obtainable magma density $M_{T,max}$ ($100*M_{solids}$ /$M_{suspension}$) at 5 °C.

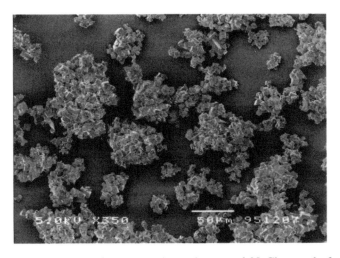

Figure 8. SEM picture of the strongly agglomerated NaCl crystals formed at $X_{DMiPA} = 0.9$ (NaCl free) and T = 5 °C.

the growth process. If this is the case there will be a depletion of sodium chloride in the mother liquor close to the crystal surface. The edges and corners of the cubic NaCl crystals have better access to the bulk liquid, where the supersaturation is much higher. At high enough bulk supersaturations surface nucleation will take place on the edges and corners, and the formed nuclei will serve as step sources. The new steps move towards the centres of the crystal faces and bunch to form macro steps. Surface nucleation on the edges and corners will continue to take place as well as the formation of new macrosteps, resulting in the formation of pyramidally shaped depressions in the faces of the crystals. Besides agglomeration the formation of these hopper crystals probably accounts for the relatively high DMiPA concentrations in the sodium chloride crystals, since mother liquor inclusions are developed under overhangs of macrosteps.

At the antisolvent fraction of 0.9 the product consists of small strongly agglomerated sodium chloride crystals. At this antisolvent fraction the crystal growth rate will be low due to the low sodium chloride solubility. This combined with the low magma density at $X_{DMiPA} = 0.9$ will result in the development of high bulk supersaturations. Extensive primary nucleation will most likely take place at the crystallizer inlets due to high local supersaturations. Primary nucleation in combination with low growth rates will result in the formation of small crystals, which tend to agglomerate in the highly supersaturated bulk solution.

At lower antisolvent fractions larger less agglomerated sodium chloride crystals are formed. This can be attributed to the fact, that the crystal growth rates are higher at lower antisolvent concentrations. Under these circumstances the crystals become larger and also the bulk supersaturation will be lower. In addition the increased magma densities at the antisolvent fractions of 0.3 and 0.6 will have a further reducing effect on the bulk supersaturation.

With DiPA continuous crystallization experiments were carried out at a DiPA fraction of 0.9 at T = 1 °C. Similarly to the experiments carried out at $X_{DMiPA} = 0.9$, the crystals formed at $X_{DiPA} = 0.9$ were strongly agglomerated. The maximum agglomerate and primary particle sizes were 200 μm and 20 μm, respectively. The DiPA concentration in the product was 590 ppm, which is comparable with the DMiPA uptake in the sodium chloride crystals produced at $X_{DMiPA} = 0.9$.

Amine Recovery by a Temperature Induced Liquid-Liquid Phase Split. To determine the feasibility of an antisolvent recovery by a temperature increase, liquid-liquid equilibria experiments were performed with amine-water-salt mixtures with compositions similar to the liquid phase compositions leaving the solid-liquid separation unit after crystallization. The liquid-liquid equilibrium lines of the ternary amine-H_2O-$NaCl_{sat}$ systems cannot be used here, because at the recovery temperature the mixtures are undersaturated with respect to sodium chloride.

Since the sodium chloride solubility at an antisolvent fraction of 0.9 is practically zero, the binary DiPA-H_2O liquid-liquid equilibrium data (see Figure 5) can be used to determine the aqueous and the organic phase compositions for the recovery of DiPA from a mixture with $X_{DiPA} = 0.9$.

Figure 9. SEM picture of the NaCl crystals formed at $X_{DMiPA} = 0.1$ (NaCl free) and T = 5 °C.

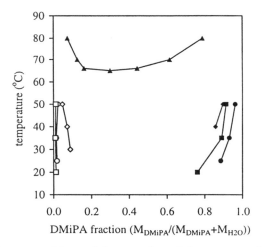

Figure 10. The compositions of the organic and the aqueous phases at several temperatures in the two liquid phase area for the DMiPA recovery from mixtures with $X_{DMiPA} = 0.1$ (●/○-marks), 0.3 (■/□-marks), and 0.6 (♦/◇-marks) saturated with NaCl at 5 °C and the binary DMiPA-H_2O L-L equilibrium line (▲-marks).

Figure 10 shows to what extent DMiPA can be recovered from spent mother liquors of X_{DMiPA} = 0.1, 0.3 and 0.6 (T_{cryst} = 5 °C) by a temperature induced liquid-liquid phase split. For the recovery of DMiPA from a spent mother liquor with X_{DMiPA} = 0.9 the binary DMiPA-H_2O liquid-liquid equilibrium data (also displayed in Figure 10) can be used to make a rough estimate of the organic and aqueous phase compositions at the recovery conditions. In Figure 10 it can be seen that the mutual solubilities of water and DMiPA decrease if more salt is present, the separation between the organic and the aqueous phases is better at higher temperatures and the DMiPA solubility in the aqueous phase is less than the water solubility in the DMiPA phase at a given temperature.

Conclusions

The amines DiPA and DMiPA are suitable candidates for the antisolvent crystallization of sodium chloride. Both amines reduce the sodium chloride solubility substantially. At the antisolvent recovery conditions the mutual solubilities of the water and the amines were low, so the separability after crystallization was generally good.

The crystallization experiments showed that the sodium chloride product consisted of small agglomerated crystals with maximum primary particle sizes of 10-70 microns. The degree of agglomeration of the crystals increased with increasing amine to brine feed ratio at which the continuous crystallization experiments were carried out. The amine uptake in the small strongly agglomerated crystals formed at an antisolvent fraction of 0.9 was approximately 600 ppm for both DiPA and DMiPA. The DMiPA concentrations in the larger less agglomerated sodium chloride crystals produced at DMiPA fractions of 0.1, 0.3 and 0.6 were below 250 ppm.

Future Work

To improve product purity, to increase crystal size and to avoid agglomeration the bulk supersaturations during crystallization should be lower and primary nucleation should be suppressed. Therefore future work will focus on crystallization experiments at higher magma densities and larger residence times. Also the importance of mixing on the antisolvent crystallization of sodium chloride will be investigated.

Acknowledgments

The authors would like to thank NOVEM for its financial support and F. v.d. Ham, E. Pentinga and M. Voorwinden for their help.

Literature Cited

1. Weingärtner, D. A.; Lynn, S.; Hanson, D.N. *Ind. Eng. Chem. Res.* **1991**, *30*, 490.
2. Davison, R. R.; Smith, W. H.; Hood, D.W. *J. Chem. Eng. Data* **1960**, *5*, 420.

Chapter 20

Semibatch Gas Antisolvent Precipitation of Poly(2,6-dimethyl-1,4-phenylene ether)

F. E. Wubbolts, R. E. A. Buijsse[1], O. S. L. Bruinsma, J. de Graauw, and G. M. van Rosmalen

Faculty of Chemical Technology and Materials Science, Laboratory for Process Equipment, Delft University of Technology, Leeghwaterstraat 44, 2628 CA, Delft, Netherlands

The polymer PPE (poly 2,6 dimethyl 1,4 phenylene ether) was precipitated from a concentrated solution in toluene using high-pressure carbon dioxide as an anti-solvent. The precipitate was characterised by the particle size, the particle size distribution, and the degree of crystallinity of the solid. The product consists of agglomerated PPE particles of approximately 1 μm. Whereas the size of these primary particles appears to be independent of the process conditions, the extent of agglomeration can be influenced by the process conditions. The temperature has the largest effect on the agglomeration process. At a relatively high process temperature of 60°C agglomerates are obtained of approximately 700 μm, with a narrow size distribution.

At high pressures a considerable amount of a gas can be dissolved in a liquid solution. Once dissolved in the liquid phase it may act as an anti solvent. In literature this type of drowning out is often referred to as the Gas Anti-Solvent (GAS) process, it is also known as the Supercritical Anti-Solvent (SAS) process or as Precipitation with a Compressed fluid Anti solvent (PCA).

Gallagher *(1)* studied the precipitation of the explosives cyclonite and nitroguanidine using high-pressure carbon dioxide as an anti-solvent to reduce solvent inclusion in the crystals. Later Chang studied the separation of β-carotene from carotene mixtures by precipitation from a liquid solution *(2)*, as well as the purification of anthracene from crude anthracene by adding compressed CO_2 *(3)*. Carbon dioxide also proved to be a suitable anti-solvent for the fractionation of a mixture of organic acids *(4)* and for the formation of microparticulate particles of insulin *(5)*.

[1]Current address: Eastman Chemical Company, Kingsport, TN 37662

Addition of an anti-solvent to a polymer solution causes the polymer solution to split into a polymer-rich phase and a solvent-rich phase. When a non-solvent is added the overall density of the original solvent becomes lower, which decreases the Lower Critical Solution Temperature (LCST) of the solution. A liquid-liquid phase-split thus occurs without raising the temperature. A low-molecular weight anti-solvent like CO_2, propane or ethane can effectively decrease the LCST of the polymer solution *(6)* and thus induce a liquid-liquid phase-split. It is due to this effect that Gas Anti-Solvent precipitation of polymers has focused on the production of polymer particles with a specific size, structure or shape such as micro tubes *(7)* or micro balloons *(8)*. Phase separation phenomena in PPE solutions during the formation of polymer membranes by the addition of a conventional anti-solvent have been described by *(9)*.

The aim of this work is to study the semi-batch precipitation of the polymer poly (2,6-dimethyl-1,4-phenylene ether) from a solution in toluene, using high-pressure CO_2 as an anti-solvent. The influence of the process conditions on the degree of crystallinity, and the particle size and size distribution has been determined.

Theory

Process conditions. In the semi-batch process an autoclave is partially filled with a solution and pressurised with CO_2. As the pressure increases, more CO_2 dissolves in the solution and the liquid level in the autoclave rises. In the binary toluene-CO_2 system the mole fraction of CO_2 in the liquid phase increases about linearly with pressure, whereas the vapour phase almost exclusively consists of CO_2 *(10)*. To determine the final pressure it is sufficient to consider the toluene-CO_2 system because hardly any PPE is left in the solution after solidification. Consequently there is no influence of the PPE on the vapour-liquid equilibrium of toluene and CO_2. To attain the same composition of the liquid phase at a higher temperature a higher pressure is required, as is shown in Figure 1. The pressures have been calculated using the Peng-Robinson equation of state in combination with the van der Waals mixing rule as was indicated by Peng *(11)*, Kwak *(12)* and Shibata *(13)*. Depending on the process temperature, pressures over 80 bar may be required to dissolve enough CO_2 in the solution to obtain solid PPE.

Plasticization of PPE by CO_2. During the precipitation of polymers it may occur that the anti-solvent is soluble in the solute. This sorption of gases and liquids in polymers can cause significant plasticization and results in a substantial decrease of the glass transition temperature, T_g *(14)*. In Figure 2 the glass transition temperature of PPE has been plotted against the CO_2 exposure pressure. The T_g of PPE appears to decrease linearly with increasing exposure pressure, which is in good agreement with theory *(15)*. By the absorption of the CO_2 the polymer first becomes rubber-like and finally melts into a fluid state. Such a mixture of PPE and CO_2 is sticky, which gives rise to another phenomenon: agglomeration.

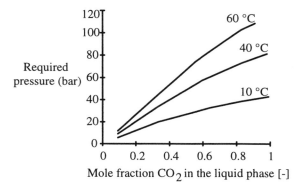

Figure 1.   Required pressure to attain a certain mole fraction of CO_2 in the liquid phase of the CO_2-toluene binary mixture.

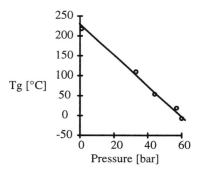

Figure 2.   The glass transition temperature of PPE as a function of CO_2 exposure pressure, measured by Hachisuka *(14)*.

Experimental

Figure 3 is a schematic diagram of the equipment that is used for precipitation, filtration and washing, all at a high pressure. It consists of a CO_2 delivery system and an autoclave with a volume of 400 ml. The vessel is equipped with baffles, a stirrer and a water-jacket for heating and cooling. The bottom of the autoclave contains a filter plate with a pore-size of 10 micrometers. The carbon dioxide flow to the precipitator is controlled by an air operated needle valve in combination with a pressure transducer and a programmable controller to impose a predetermined pressure profile.

Experimental procedure. The autoclave is partially filled with a clear solution of PPE in toluene and closed. The pressure is raised linearly from atmospheric to the desired final pressure by supplying carbon dioxide to the bottom of the precipitator (V-1 open). The CO_2 bubbles through the solution, where it is absorbed by the liquid phase. Close to the final pressure the liquid level begins to rise rapidly and the clear solution turns turbid white. After the final pressure has been reached the solution is held at this pressure until equilibrium is attained and the autoclave is just filled with liquid. During this equilibration period CO_2 must be added to maintain the pressure because it dissolves in the liquid phase. Generally it takes 30 minutes before the addition of CO_2 stops. Because the gaseous CO_2 'condenses' into the liquid, heat is released during dissolution. As a result the temperature in the vessel rises approximately 5 to 10°C, depending on the rate at which the CO_2 dissolves and heat can be removed from the vessel by the water jacket.

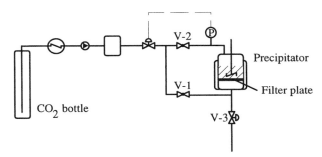

Figure 3.   Schematic drawing of the experimental set-up used for the batch experiments

After equilibrium is reached the needle valve (V-3) is opened to separate the solid PPE from the mother liquor. The mother liquor leaves through the bottom of the autoclave while the PPE remains on the filter plate. During this operation the pressure in the autoclave is kept constant by allowing the CO_2 to enter from the top of the precipitator and by closing the supply-route through the bottom. To remove the adhering mother liquor the precipitate is washed thoroughly with 1 kg of CO_2 at the same pressure for around 60 minutes. If the precipitate is not washed adequately the remaining mother liquor will regain its solvent power upon depressurisation.

Process parameters. Within the experimental procedure a number of parameters can be varied. In a series of preliminary experiments, a suitable standard value was chosen for the temperature (20°C), the pressure build-up rate (2 bar/min), the stirrer rate, the initial relative concentration of the polymer solution and the relative liquid phase expansion. As the solubility of PPE in toluene increases with increasing temperature a relative concentration has been introduced which is defined as the ratio of the polymer concentration in the initial solution and the saturation concentration at the process temperature. All parameters were varied separately around this standard value, as is indicated in Table I, to determine their individual effect on the particles' size and shape.

At the end of an experiment the liquid just fills the autoclave. The final CO_2 : toluene ratio in the vessel was varied by changing the amount of solution that is initially charged into the vessel. To make sure that the autoclave was just filled with liquid when vapour-liquid equilibrium is attained at the end of an experiment, the final pressure was calculated beforehand using the Peng-Robinson equation of state. A good agreement between theory and experiment was observed. The final pressure that has to be attained depends on the temperature and the final CO_2 : toluene ratio and is therefore not regarded as a variable in the process. As the CO_2 dissolves the liquid level rises. The solution undergoes an 'expansion' which is related to the initial amount of solution and the volume of the autoclave. The expansion of the solution is taken as a measure for the ratio CO_2 : toluene.

Table I.   The process conditions during the experiments

		Standard experiment		
Relative liquid phase expansion (-)		1	3.5	6	
Relative concentration PPE (-)	0.1	0.25	0.5	0.75	0.9
Pressure build-up rate (bar/min)		0.5	2	100	
Temperature (°C)		10	20	40	60
Stirrer rate (RPM)	50	150	300	600	900

$$\frac{\Delta V}{V} = \frac{V_{prec} - V_{sol}}{V_{sol}}$$

where: $\Delta V/V$ is the expansion of the liquid phase (-)

  V_{prec} is the volume of the precipitator (ml)

  V_{sol} is the volume of the initial solution (ml)

Other experiments. In the same semi-batch equipment one experiment was performed using methanol to precipitate the PPE instead of CO_2. In another experiment a Styrene-Butadiene-Styrene (SBS) triblock copolymer was precipitated from cyclohexane by adding CO_2 as an anti-solvent. This synthetic rubber contains blocks with a glass transition temperature of 100°C (polystyrene) and blocks with a glass transition temperature of approximately -100°C (polybutadiene).

Analyses. Because of the high contrast and magnification factor a Scanning Electron Microscope (SEM) was used to study the surface morphology of the precipitate. The particle size and size distribution were determined by laser light scattering (Coulter LS). Thermal properties were determined by Differential Scanning Calorimetry (DSC) to establish the weight fraction of crystalline material in the solid PPE. These measurements were confirmed by X-ray diffraction.

Results and Discussion

The PPE was obtained from the precipitator as a free flowing powder. Filtration and washing of the product need much attention because the adhering mother-liquor regains its solvent power upon depressurisation. If the filtration and washing are performed improperly, the precipitated particles glue together and form gel-like lumps that contain much of the original solvent. Due to the development of a good experimental procedure all experiments were very well reproducible.

Figure 4 shows that this powder consists of agglomerates which are composed of spherical PPE particles of approximately 1 μm. Whereas the size and shape of the primary particles are independent of the varied process conditions the extent of agglomeration of those primary particles can be controlled by changing the process conditions. At the selected standard process conditions, which are indicated in table (I), a product is obtained with an average agglomerate size of 35 μm. More rapid addition of CO_2, at a higher pressure build-up rate, reduces the extent of agglomeration and the average size of the particles drops to 10 μm. By increasing the process temperature from its standard value of 20°C to 60°C, the size of the agglomerates increases from 20 μm to over 600 μm. Within the range studied the stirrer rate, the expansion and the relative concentration of PPE have little to no effect on the size of the agglomerates.

Analysis of the product with DSC showed that in most samples the degree of crystallinity was in the order of 25 wt %. Only at a high pressure build-up rate the amount of crystalline material decreased to 17 wt %.

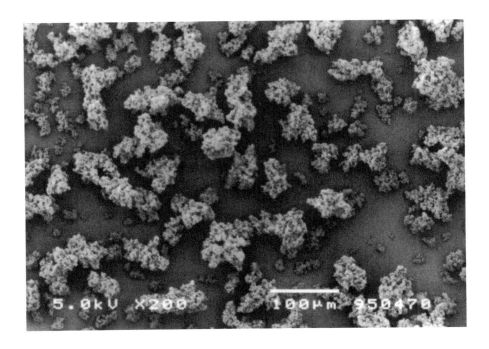

Figure 4.   SEM picture of the product as it is obtained at the standard process conditions, table (I).

Mechanism of Solidification. Addition of the non-solvent lowers the LCST of the polymer solution. When enough anti-solvent is added the LCST of the solution drops below the process temperature and a polymer-rich phase emerges that is finely dispersed in a solvent-rich phase. Due to the high concentration of the polymer in the concentrated phase, the polymer quickly solidifies and the residual solvent is taken up by the solvent-rich phase. We assume that this is the most likely mechanism for the creation of the primary particles that subsequently form the agglomerates.

The influence of the pressure build-up rate. Comparison of the product in Figure 4 and in Figure (6, above) shows that rapid addition of the CO_2 results in small agglomerates. In these smaller agglomerates the primary particles can clearly be distinguished and the agglomeration appears to be looser than in the larger agglomerates which are obtained at the standard process conditions. Rapid addition of the CO_2 also results in a wider distribution of the particles' size, as is shown in Figure 5. The reason for both effects is, that it takes less time to reach the final pressure at a high pressure build-up rate. Therefore the overall process time is much shorter and less time is available for agglomeration of the particles.

Slow addition of the CO_2 results in somewhat larger agglomerates of 45 μm. The surface of those particles is clearly flattened by the stirrer and the agglomerates appear to be denser than in the product that was formed at the standard process conditions. Both the flattening of the surface and the somewhat larger agglomerates are a result of the longer process time.

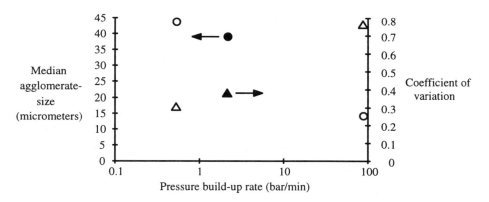

Figure 5.   The influence of the pressure build-up rate on the agglomerate size. The ● and ▲ indicate the standard experiment.

Also the decrease of the degree of crystallinity at high pressure build-up rates can be explained by the shorter process time. Polymers do not crystallise in the same way as most other classes of compounds do: usually the crystalline regions are small and crystallisation proceeds very slowly, due to the large size of the polymer molecules. The length of the polymer chains hampers their orientation in a crystalline structure because the molecules are entangled. A large part of the polymer therefore

Figure 6.   SEM pictures of the particles formed a high pressure build-up rate of 100 bar/min (above) and at a high process temperature of 60°C (below) at the same magnification.

solidifies in an amorphous state, where the chains are not arranged in an ordered structure but have assumed random orientations. To obtain a solid polymer with a relatively high degree of crystallinity the molecules must be allowed time to orientate themselves. Presumably the opportunity for orientation of the polymer molecules during the solidification in the polymer-rich phase is limited by the process time, when the CO_2 is rapidly added to the solution.

The influence of the temperature on the extent of agglomeration. The SEM pictures in the Figures 4 and (6,below) show the PPE agglomerates that are formed at the standard process conditions and at a temperature of 60°C respectively. By increasing the temperature from 20 to 60°C the size of the agglomerates increases from 40 to over 600 µm, as is shown in Figure 7. The corresponding pressure (Figure 1) plays an important role in the explanation of the increase of the agglomerate size at higher temperatures.

To dissolve enough CO_2 in the solution and obtain solid polymer at a higher process temperature, the final pressure in the autoclave must be higher as can be seen from Figure 1. Although it takes more time to attain the final pressure, because the pressure build-up rate remains the same, this is not the reason for the increase of the agglomerate size.

The influence of the CO_2 on the properties of the PPE may not be neglected. When the pressure is high enough the CO_2 dissolves in the PPE and acts as a plasticizer, which causes an appreciable decrease in the glass-transition temperature of the polymer. The effect of the CO_2 exposure pressure on the glass transition temperature of PPE has been indicated in Figure 2. Pure PPE is a solid material with a T_g of 220°C and at a process temperature of 60°C PPE is expected to be a solid material. At high pressures, however, CO_2 dissolves in the amorphous PPE and 'expands' the amorphous polymer, which increases the mobility of the polymer chains. The glass transition temperature of the PPE-CO_2 mixture is therefore lower than the T_g of pure PPE, and the material becomes weak and sticky at a much lower temperature.

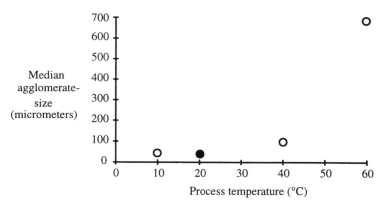

Figure 7.   The influence of the process temperature on the size of the agglomerates. The • indicates the standard experiment.

An increase in the process temperature is thus accompanied by a decrease of the polymer's glass transition temperature. Figure 8 shows the relation between the glass transition temperature of the PPE and the pressure, as well as the operating range of the GAS process. Figure 7 shows that the extent of agglomeration clearly increases at temperatures higher than 40°C, which corresponds to the area where the line indicating the T_g enters the operating range of the GAS process in Figure 8.

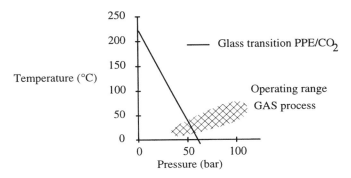

Figure 8.   The process- and glass transition temperature as a function of the pressure.

Other experiments. Precipitation of PPE from toluene with methanol yielded the same kind of agglomerates as with CO_2. A major difference between the experiments was that only a few drops of methanol had to be added to turn the clear polymer solution turbid white, whereas when CO_2 is used the liquid level first rises considerably. The way in which the solubility declines is therefore different for methanol and CO_2. In both cases the amount of crystalline material was approximately 25 %, and does not change significantly when methanol is used instead of CO_2.

During the experiment with precipitation of the polymer SBS from a solution in cyclohexane the solution did not turn turbid white when the CO_2 was added. Therefore it first seemed that no material had precipitated. However, during depressurisation the inside of the autoclave slowly turned white and when the vessel was opened it appeared that the internals and the vessel's wall had been coated with a thin foam layer of approximately 2 mm of polymer. The layer contained a large number of small bubbles, presumably CO_2 that was trapped in the polymer. Apparently a polymer-rich phase was formed during the addition of the CO_2. This viscous phase was flung to the wall and the internals by the stirrer, where it remained behind when the autoclave was flushed with CO_2 and the solvent-rich phase was withdrawn. During depressurisation the CO_2 and residual cyclohexane evaporated and a thin foam layer covered the interior of the vessel.

Conclusions

It is important to note that the experiments were very well reproducible. This is mainly due to the fact that a good procedure has been developed for filtration and washing of the product after precipitation. Filtration and washing need much attention because the adhering mother-liquor regains its solvent power upon depressurisation.

As the anti-solvent is added the polymer solution splits into a polymer-rich phase and a solvent-rich phase. In the highly concentrated polymer-rich phase the PPE solidifies and the residual solvent is taken up by the solvent-rich phase. The size of these solid PPE particles is approximately 1μm and independent of the varied process conditions.

Whereas the size of the primary particles is independent of the varied process conditions, their extent of agglomeration can be influenced by the rate of anti-solvent addition and the process temperature. By quickly adding the anti-solvent, at a high pressure build-up rate, the extent of agglomeration is reduced because the process time becomes shorter and less time is available for agglomeration.

At a relatively high process temperature a high pressure is required to dissolve enough CO_2 in the solution to obtain solid PPE. However, at a higher pressure the CO_2 also dissolves in the PPE and lowers its glass-transition temperature below the process temperature. The material thus becomes sticky and large agglomerates are formed.

Literature Cited

1. Gallagher P.M., Coffey M.P., Krukonis V.J., Klasutis N., Gas Anti-Solvent recrystallization: new process to recrystallize compounds insoluble in supercritical fluids, *Supercritical fluid science and technology,* Johnston K.P., Penninger J.M.L., **1989**, *(406)*, 334-354, .
2. Chang C.J., Randolph A.D., Separation of beta-carotene mixtures precipitated from liquid solvents with high pressure CO_2, *Biotechn. Prog.* **1993**, *7(3)*, 275-278
3. Liou Y., Chang C., Separation of Crude Anthracene Using Gas Anti-Solvent Recrystallization, *Separation Science and Technology*, **1992**, *(27-10)*, 1277-1289
4. Shishikura A., Kanamori K., Takahashi H., Kinbara H., Separation and purification of Organic Acids by Gas Anti-Solvent Crystallisation, *J. Agric. Food Chem.* **1994**, *(42)*, 1993-1997
5. Yeo S.D., Lim G.B., Debenedetti P.G., Bernstein H., Formation of microparticulate protein powders using a Supercritical Fluid Antisolvent, *Biotechn. and Bioeng.* **1993**, *(41)*, 341-346
6. Irani C.A., Cozewith C., Lower critical solution temperature behaviour of ethylene propylene copoloymers in multicomponent solvents, *J. of Applied Polymer Science*, **1986**, *(31)*, 1879-1899
7. Luna-Bárcenas G., Kanakia S.K., Sanchez I.C., Johnston K.P., Semicrystalline microfibrils and hollow fibres by precipitation with a compressed-fluid antisolvent, *POLYMER*, **1995**, *(36-16)*, 3173-3182
8. Dixon D.J., Luna-Bárcenas G., Johnston K.P., Microcellular microspheres and microballoons by precipitation with a compressed fluid anti-solvent, *POLYMER*, **1994**, *(35-18)*, 3998-4005
9. Wijmans J.G., Rutten H.J.J., Smolders C.A., Phase separation phenomena in solutions of Poly(2,6-dimethyl-1,4-phenylene Oxide) in mixtures of trichloroethylene, 1-octanol and methanol: Relationship to membrane formation, *Journal of Polymer Science: polymer physics edition*, **1985**, *(23)*, 1941-1955

10. Ng H.J., Robinson D.B., Equilibrium phase properties of the toluene-carbon dioxide system, *Journal of Chemical and Engineering data*, **1978**, *23(4)*, 325-327

11. Peng, D.Y., Robinson D.B., A new two constant equation of state. *Ind. Eng. Chem. Fundam.*, **1976**, *15(1)*, 59-64

12. Kwak, T.Y., Mansoori G.A., Van der Waals mixing rules for cubic equations of state: Applications for supercritical fluid extraction modelling. *Chem. Eng. Sci.*, **1986**, *41(5)*, 1303-1309

13. Shibata S.K., Sandler S.I., Critical evaluation of equation of state mixing rules for the prediction of high-pressure vapour-liquid equilibria, *Ind. Eng. Chem. Res.* **1989**, *(28)*, 1893-1898

14. Hachisuka, H., T. Sato, T. Imai, Y. Tsujita, A. Takizawa and T. Kinoshita, Glass transition temperature of glassy polymers plasticized by CO_2 gas, *Polymer Journal*, **1990**, *(22-1)*, 77-79

15. Condo P.D., Sanchez I.D., Panayiotou C.G., Johnston K.P., Glass transition behaviour including retrogade vitrification of polymers with compressed fluid diluents, *Macromolecules* **1992**, *(25)* 6119-6127,

Chapter 21

Reactive Crystallization of Magnesium Hydroxide

Hideki Tsuge, Kyoichi Okada, Toru Yano, Naoko Fukushi,
and Hiroko Akita

Department of Applied Chemistry, Keio University, 3–14–1 Hiyoshi,
Kohoku-ku, Yokohama 223, Japan

Magnesium hydroxide was produced from magnesium chloride and alkali feed by using a CMSMPR crystallizer to clarify the characteristics of reactive crystallization kinetics of magnesium hydroxide. The following factors were investigated affecting the reactive crystallization kinetics; the nature of alkali feeds, the mole fraction of NaOH in the mixture of $Ca(OH)_2$ and NaOH, and additives. The growth rate G and the birth rate B^o are correlated by the power law model represented by $B^o \propto G^i$. The kinetic order i in the power law model depends on the system of reaction.

Some valuable natural resources dissolved in the sea water, such as sodium chloride, magnesium salts and bromine, are used in the chemical industries. Especially, magnesium salts are produced from magnesium hydroxide obtained by the reactive crystallization of the sea water and calcium hydroxide in the salt industries. Magnesium hydroxide has been used industrially as desulfurization agent and as construction material of steel plant refractory. Also it is used for the production of basic magnesium carbonate, whose dominant particle size depends on the form and the dominant particle size of the raw material, magnesium hydroxide. So it is necessary to produce the most suitable raw particle of magnesium hydroxide in form and in particle size.

Several studies have been done on the reactive crystallization of magnesium hydroxide. Murotani et al.(1) precipitated magnesium hydroxide by adding ammonia to mixed solution of magnesium chloride and ammonium chloride. In this reaction system hexagonal crystals of magnesium hydroxide were obtained by adding sodium hydroxide as additive. Phillips et al.(2) precipitated magnesium hydroxide at 60 ℃ at various constant pH levels (8.7 ~ 12.5) from magnesium chloride and ammonium(or sodium) hydroxide. They indicated that excess OH^- or Mg^{2+} has different effects upon the growth rate in a particular crystallsgraphic direction, which leads to changes in the shape and the diameter to thickness ratio of particles. Dabir et al.(3) studied the precipitation kinetics of magnesium hydroxide obtained from aqueous solutions of magnesium chloride and sodium hydroxide by using a CMSMPR crystallizer. They correlated the kinetic order i with hydroxide concentration, although their experimental ranges were rather narrow. Mullin et al.(4) carried out precipitation of magnesium hydroxide by mixing aqueous solutions of magnesium chloride and sodium hydroxide. They discussed the aging of precipitated magnesium hydroxide and indicated that magnesium hydroxide consisting of primary particles(~ 40nm) agglomerated into larger particles(~ 20 μ m). Tsuge and Matsuo(5) measured the crystallization kinetics of primary magnesium hydroxide particles produced by magnesium chloride and calcium hydroxide in a CMSMPR crystallizer and correlated the kinetic order i with concentrations of hydroxide and chloride ions. Tsuge et al.(6) also measured the crystallization kinetics of agglomerated magnesium hydroxide particles from magnesium chloride, magnesium sulfate and calcium hydroxide. And Tsuge et al.(7) crystallized agglomerated magnesium hydroxide particles by mixing aqueous solution of calcium hydroxide and artificial sea water, and correlated the kinetic order i with chloride ion concentration or ion strength. Turek(8) tested precipitation of magnesium hydroxide from hard coal mine brine and sodium hydroxide.

It is essential to clarify the crystallization kinetics of magnesium hydroxide and to investigate the influence of various factors on the reactive crystallization characteristics of magnesium hydroxide for better understanding of its crystallization process. The aims of this study are to discuss the followings;

1)the effect of alkali feed on the crystallization kinetics of magnesium hydroxide,
2)the effect of mole fraction of NaOH in the mixture of Ca(OH)$_2$ and NaOH, $R(=[OH^-]_{NaOH}/([OH^-]_{Ca(OH)_2}+[OH^-]_{NaOH}))$ on the agglomeration process of magnesium hydroxide,
3)the effect of the additive on the dominant particle size of magnesium hydroxide,
4)the effect of the reaction system on the kinetic order i in the power law model.

Experimental

Experimental apparatus and procedure.   **Figure 1** shows a schematic diagram of the experimental apparatus. A 1 liter stirred tank reactor made of acrylic resin was used as a CMSMPR crystallizer. The impeller used was of 6–blade turbine and operated at 450 rpm. Feed solutions were pumped continuously into the crystallizer to produce magnesium hydroxide. The reaction temperature was maintained at 25 ℃ .

Table I . Experimental Conditions of Series 1

		Initial Concentration [mol/m^3]
	$MgCl_2$	2.0, 4.0, 8.0
Alkali feed	NaOH, KOH	4.0, 8.0, 16.0
	$Ca(OH)_2$, $Ba(OH)_2$	2.0, 4.0, 8.0

Table II . Experimental Conditions of Series 2

Initial concentration of Mg^{2+} and OH^-	2.0, 4.0, 8.0 mol/m^3
Mole fraction of NaOH in alkali feed R	0, 0.3, 0.6, 1.0

$*R = [OH^-]_{NaOH}/([OH^-]_{Ca(OH)_2} + [OH^-]_{NaOH})$

Table III −1. Experimental Conditions of Series 3

Initial concentration of $Ca(OH)_2$	2.0 mol/m^3
Initial concentration of $MgCl_2$	2.0 mol/m^3
Concentration of NaCl	0, 18.8, 36.7, 73.8, 95.1 mol/m^3
Concentration of $CaCl_2$	0, 10.0, 20.0, 40.0 mol/m^3

Table III −2. Experimental Conditions of Series 3

Initial concentration of NaOH	4.0 mol/m^3
Initial concentration of $MgCl_2$	2.0 mol/m^3
Concentration of $CaCl_2$	0, 20.0, 40.0, 80.0 mol/m^3

Sampling of the suspension was begun after 10 residence times, when the steady state had been reached. After the filtration and natural drying, the particles obtained were analysed by x-ray diffraction, and phtographed by scanning electron microscope (SEM) and their sizes were analyzed by using a digitizer. Irrespective of crystal form, the maximum length of primary and agglomerated particles was used as the crystal size.

Experimental conditions.   All experiments were conducted with stoichiometric feed ratio and magnesium chloride was used as the magnesium source. The residence times of reactants in the crystallizer were 300, 600 and 1200 seconds.

The following three experimental series were studied;
1)The influence of alkali feed on the behavior of the crystallization of magnesium hydroxide was investigated. As alkali feeds, sodium hydroxide, potassium hydroxide, calcium hydroxide and barium hydroxide were used. Experimental conditions of series 1, that is, initial concentrations of magnesium chloride and alkali feeds, are listed in **Table I** .
2)From the experimental results in series 1, it was found that the particle properties of magnesium hydroxide depend on the type of alkali feed. To change the dominant particle sizes of primary and agglomerated particles, it may be effective to use the mixture of alkali feeds. In series 2, calcium hydroxide and sodium hydroxide were used as two kinds of alkali feed. Experimental conditions of series 2 are listed in **Table II** , where R means the mole fraction of sodium hydroxide in the mixture of calcium hydroxide and sodium hydroxide, that is, $[OH^-]_{NaOH}/([OH^-]_{Ca(OH)_2}+[OH^-]_{NaOH})$.
3)In general, particle size is reduced by additives. To investigate the effect of additives, sodium chloride or calcium chloride was added in the reaction system in which calcium hydroxide or sodium hydroxide was used as alkali feed. Experimental conditions of series 3 are listed in **Tables III −1 and III −2.**

Analysis of the PSD Data.   From the population balance for a CMSMPR crystallizer operated under steady−state condition, the population density n for size−independent crystal growth is given by Equation (1), where $n^°$, G, θ and l are nuclei density, growth rate, residence time of reactants and particle size, respectively.

$$n = n^° \exp\left(-\frac{l}{G\theta}\right) \tag{1}$$

Figure 2 shows the typical particle size distribution(PSD) of magnesium hydroxide plotted on semilogarithmic coordinates. The linear correlation indicates that the primary particle growth obeys the Δ L law. From the slope of PSD, G is

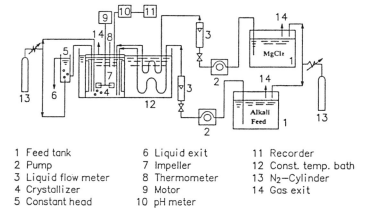

1 Feed tank	6 Liquid exit	11 Recorder
2 Pump	7 Impeller	12 Const. temp. bath
3 Liquid flow meter	8 Thermometer	13 N_2–Cylinder
4 Crystallizer	9 Motor	14 Gas exit
5 Constant head	10 pH meter	

Figure 1. Experimental apparatus

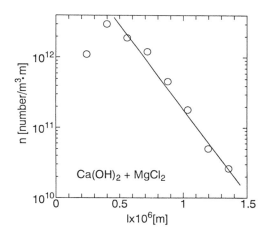

Figure 2. Particle size distribution of Mg(OH)$_2$

obtained. And the dominant particle size l_m and the birth rate B° are calculated from Equations (2) and (3).

$$l_m = 3G\theta \tag{2}$$

$$B^\circ = \frac{9P_r}{2l_m^3 \rho_c f_v V} \tag{3}$$

$$P_r = (C_{oMg2+} - C_{Mg2+})\, F \tag{4}$$

The production rate P_r in Equation (3) is calculated from Equation (4) from mass balance of magnesium ion, where M, C_{oMg2+}, C_{Mg2+} and F are molecular weight, initial concentration of Mg^{2+}, concentration of Mg^{2+} in the crystallizer and feed rate, respectively.

By plotting the particle size distribution(PSD) of agglomerated particles on semilogarithmic coordinates, a linear correlation was obtained, which has the same correlation coefficient with that in primary particle. PSD data of agglomerated particles was analysed by assuming that the particle growth process in agglomerated particles is similar to that in primary particles, whereas the growth unit in agglomerated particles is much bigger than that in primary particles. The size–independent particle growth may be applied to the agglomerated particle growth, and B°, G and l_m are calculated in the same way as in primary particles, where B° means the apparent birth rate for agglomerated particle.

Results and Discussion

Observation of magnesium hydroxide particles by SEM photographs.
Figure 3 shows SEM photographs of magnesium hydroxide particles which were produced from magnesium chloride and various alkali feeds. The crystal structure of magnesium hydroxide is CdI_2 type and the crystal form of primary particle is a disk. From these SEM photographs, primary particles were observed mainly when calcium hydroxide was used as alkali feed(the system of calcium hydroxide), whereas agglomerated particles which were consisted of many larger primary particles were observed when other alkalis were used as feed(other alkali feed systems). In the system of calcium hydroxide agglomerated particles were also observed by SEM photograph, whereas they were much smaller in size of primary particle and of lower frequency of agglomeration than those in other alkali feed systems.

Effect of alkali feed.
Figure 4 shows the relationship between growth rate G and residence time θ. Growth rate decreases with increasing residence time because of decrease in supersaturation. SEM photographs show that the particle growth depends mainly on the primary particle growth in calcium hydroxide feed and the agglomeration of primary particles in other alkali feed systems. As the

$$C_{oMgCl_2} = 4mol/m^3$$

$$\theta = 1200s$$

Figure 3. SEM photographs of Mg(OH)$_2$

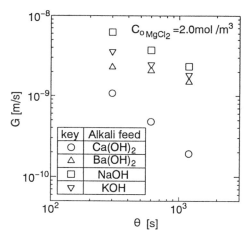

Figure 4. Relation between G and θ

growth unit of agglomerated particle is bigger than that of primary particle, the growth rates of agglomerated particles are bigger than that of primary particles and the slopes of other alkali feed systems are much more moderate than that of system of calcium hydroxide as shown in **Figure 4**.

Figure 5 shows the relationship between dominant particle size l_m and residence time θ. Dominant particle size decreases slightly with increasing residence time for the system of calcium hydroxide, whereas dominant particle size increases with increasing residence time in other alkali feed systems. This tendency corresponds to the difference in the inclination of $G-\theta$ curve.

The solubility and hydration number of alkali feed, diffusion coefficient of reacted ions and interaction of ion pair affect the reactive crystallization. As a result of investigation of a number of relationships between these properties and crystallization characteristics, the key property of alkali feeds may be considered as the following.

Table IV shows the solubilities of various alkali feeds. In general, smaller solubility means that the movement of OH^- is controlled by ion pair of solute. So it may be difficult for Mg^{2+} to collide with OH^- in calcium hydroxide which has the smallest solubility. From SEM photographs of magnesium hydroxide in **Figure 3**, primary particles are formed in the system of calcium hydroxide, whereas larger primary particles are formed and agglomerated easily in other alkali feed systems which have larger solubility. Then dominant particle size of agglomerated particle becomes larger. Therefore, the solubility may be the key property among the above−mentioned properties. This tendency was also observed in the reactive crystallizations of other carbonates, that is, calcium carbonate(9) and lithium carbonate(10).

Effect of mole fraction of NaOH in the mixture of Ca(OH)$_2$ and NaOH, R.

Figure 6 shows SEM photographs of magnesium hydroxide particles and indicates that the ratio of agglomerated particle to primary particle increases with increase in mole fraction of NaOH in the mixture of Ca(OH)$_2$ and NaOH, R.

Figure 7 shows the relationship between l_m and R. The dominant particle size increases with increasing R. By using two kinds of alkali feeds which have different solubilities, the size of primary particle and the frequency of agglomeration may be changed. Then the size of agglomerated particle may be controlled by R.

Effect of additive.

Figure 8 shows the relationship between dominant particle size l_m and concentration of additive C_A in each reaction system. The dominant particle size decreases in all cases of addition of chloride. But the concentration of additive does not affect the on dominant particle size to the same extent all systems. The difference of the dominant particle size is due to the alkali feed. Primary particles are agglomerated easily in the system of sodium hydroxide, whereas they are hardly agglomerated in the system of calcium hydroxide.

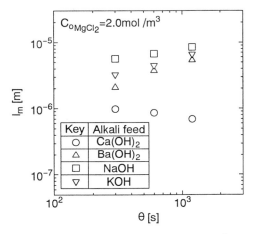

Figure 5. Relation between l_m and θ

Table IV .   Solubilities of Various Alkali Feeds at 25 ℃

Alkali feed	Solubility[g/100g H_2O]
Ca(OH)$_2$	0.129
Ba(OH)$_2$	2.393
KOH	27.24
NaOH	31.54

$$C_{oMgCl_2} = 4\text{mol}/\text{m}^3$$

$$\theta = 1200\text{s}$$

Figure 6. SEM photographs of Mg(OH)$_2$

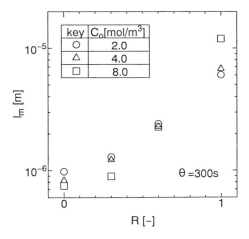

Figure 7. Relation between l$_m$ and R

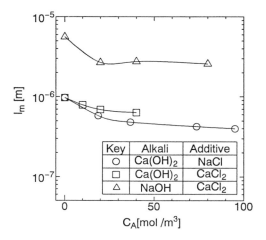

Figure 8. Relation between l $_m$ and additive concentration

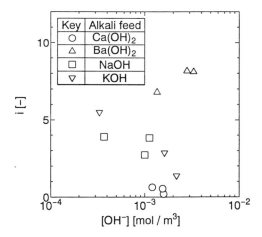

Figure 9. Relation between i and [OH⁻]

In calcium hydroxide, Na^+ and Ca^{2+} hardly influence the dominant particle size. Then it is considered that the crystallization kinetics of magnesium hydroxide does not depend on the cation in the crystallizer, but on the properties of each alkali feed as shown in **Figure 5**.

Kinetic order i. The relationship between birth rate and growth rate is written by Equation (5) and i is the kinetic order.

$$B^\circ = kG^i \tag{5}$$

Figure 9 shows the relationship between the kinetic order i and the concentration of hydroxide ion in the crystallizer. The kinetic order i is less than 1 in the system of calcium hydroxide, whereas i is larger than 1 in other alkali feed systems. This phenomenon is explained as follows: In the systems of alkali feeds except calcium hydroxide, birth rate is influenced more greatly than growth rate by residence time. Then the gradient of B° -G curve is steep, hence i becomes large.

Figure 10 shows the relationship between the kinetic order i and the concentration of hydroxide ion in the crystallizer in the systems of mixture of alkali feed and additive. The kinetic order i is less than 1 when calcium hydroxide exists in the crystallizer.

Conclusions

Reactive crystallization experiments of magnesium hydroxide were conducted to clarify the characteristics of reactive crystallization kinetics by the CMSMPR crystallizer. The following conclusions were obtained;
1) Alkali feed influences the properties of magnesium hydroxide particles. In the system of calcium hydroxide primary particles are formed easily and the kinetic order i is less than 1. But, in other cases, that is, barium hydroxide, potassium hydroxide and sodium hydroide as alkali feeds, agglomerated particles are formed easily and kinetic order i is greater than 1.
2)The dominant particle size may be controlled by the mole fraction of NaOH in the mixture of $Ca(OH)_2$ and NaOH, and additive. So the dominant particle size may be controlled by the mixture of alkali feeds, which are different in solubility.
3)The dominant particle size decreases slightly by the addition of additives.

Legend of symbols

B°	birth rate	$m^{-3} s^{-1}$
C	concentration	mol/m^3
C_A	concentration of additive	mol/m^3
C_0	initial concentration	mol/m^3
f_v	volume shape factor	—
F	feed rate	m^3/s

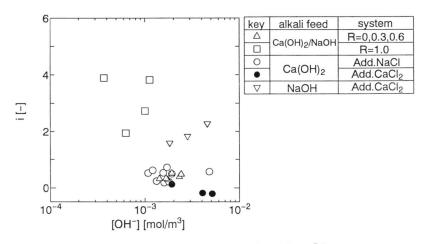

Figure 10. Relation between i and [OH⁻]

G	growth rate	m/s
i	kinetic order	–
l	particle size	m
l $_m$	dominant particle size	m
M	molecular weight	kg/mol
M $_T$	suspension density	kg/m^3
n	population density	m^{-3} m^{-1}
n^o	nuclei density	m^{-3} m^{-1}
Pr	production rate	kg/s
V	volume of crystallizer	m^3
θ	residence time	s
ρ_c	crystal density	kg/m^3

Literature Cited

1. Murotani,H and T.Shirasaki, *Bull.Tokyo Inst.Technol.*1961,42,15–26.
2. Phillips,V.A.,J.L.Kolbe and H.Opperhauser, *J.Crystal Growth*.1977,41,228–234.
3. Dabir,B.,R.W.Peters and J.D.Stevens, *Ind.Eng.Chem.Fundam*.1982,21,298–305.
4. Mullin.J.W.,J.D.Murphy et al., *Ind.Eng.Chem.Res*.1989,28,1725–1730.
5. Tsuge,H. and H.Matsuo, *ACS Symp.Series*,1990,438,344–354.
6. Tsuge,H.,Y.Kotaki and K.Terada, *11th Symp. on Ind.Crystallization*, 1990, 223–228.
7. Tsuge,H.,Y.Kotaki and S.Asano, *7th Symp.on Salt*,1993, Ⅱ ,219–223.
8. Turek,M. and W.Gnot, *Ind.Eng.Chem.Res*.1995,34,244–250.
9. Kotaki,Y. and H.Tsuge, *J.Crystal Growth*,1990,99,1092–1097.
10.Tsuge,H.,H.Yokouchi and Y.Kotaki, *Proc.Ind.Crystallization'93*.1993,2,141–146.

Chapter 22

Phosphate Recovery by Reactive Crystallization of Magnesium Ammonium Phosphate: Application to Wastewater

Izumi Hirasawa, Hiroyuki Nakagawa, Osamu Yosikawa, and Masanori Itoh

Department of Applied Chemistry, Waseda University, 3–4–1 Okubo, Shinjuku-ku, Tokyo 169, Japan

In this study, reactive crystallization of magnesium ammonium phosphate (MAP) has been studied experimentally in a batch agitated crystallizer. Operating conditions for selective crystallization of MAP were obtained. Under these conditions, batch crystallization experiments were performed, to obtain the behavior of concentration decrease in phosphate and magnesium ions, by changing the initial Mg/P ratio. It was recognized that the concentration decrease pattern affected the properties of produced crystals. A new process for removing ammonium and phosphate ions using a crystallization process was proposed for application to wastewater from sludge treatment processes.

The authors have already reported a new phosphate removal process by calcium phosphate crystallization[1]. However in order to cope with the eutrofication problem, removal of ammonium ion should also be considered. For the removal of ammonium ions, a biological denitrification process was established, but that process had problems needing much installation area for the equipment and not to be able to recover ammonium ion. On the other hand, sludge treatment equipment, to treat sludge in a concentrated way, is planned in the large city of Tokyo, as that city doesn't have much space for the sludge treatment. If such sludge treatment will be in practice, large amounts of wastewater containing ammonium and phosphate ions will be produce during the solid-liquid separation of sludge.

We experienced a scaling problem(which consisted mainly of MAP) in the pipeline of the digestion tank in the sewage treatment process. The reason of scaling was considered to be that the solution in the tank gained supersaturation in MAP caused by the release of ammonium and phosphate ions from the sludge in the process of digestion[2]. MAP crystals are known to be produced by the reaction in Eq. (1).

$$Mg^{2+} + NH_4^+ + HPO_4^{2-} + OH^- + 6H_2O \rightarrow MgNH_4PO_4 \cdot 6H_2O + H_2O \qquad (1)$$

Focusing on these phenomena, the application of MAP crystallization to the wastewater of sludge treatment was considered. The merit of this process is that ammonium and phosphate ions could be removed and recovered in the form of a slow release fertilizer.

In this paper, the reactive crystallization of MAP has been studied in a batch agitated crystallizer, in order to obtain the crystallization characteristics and the effect of Mg/P molar ratio on the properties of produced crystals and the behavior of the concentration decrease.

Experiments

Experiments were performed by using a batch reactive crystallization apparatus as shown in Fig.1, in order to obtain the properties of the produced crystals in the Mg^{2+}-NH_4^+-PO_4^{3-}-OH-H_2O system, the effect of pH on the solubility of MAP, and the effect of Mg/P molar ratio on the behavior of the concentration decrease and the properties of crystals.

Phosphate concentration was measured by the molybdenum blue absorptiometry, and magnesium and ammonium ions were measured by cation chromatography. Properties of crystals were analyzed by Xray analysis and SEM photographs.

(1) Crystallization characteristics of MAP

Crystallization experiments were performed by varying temperature, pH and concentrations of the reactants : phosphoric acid, ammonium chloride and magnesium chloride. The pH was adjusted by sodium hydroxide solution.

Mixed suspension was sampled, when the concentration of each reactant was observed to be constant. After that, the suspension was filtered through 0.45 μ m membrane filter, to be analyzed by Xray diffractometer.

(2) Effect of pH on the solubility of MAP

Based on the results of experiment (1), crystallization experiments were done in the presence of excess MAP crystals, changing the pH conditions as shown in Table 1, to obtain the concentration of the reactant after 24h. Also the phosphate removal efficiency was obtained under the same experimental condition , changing pH and Mg/P molar ratio.

(3) Effect of Mg/P molar ratio on the behavior of concentration decrease and the properties of produced crystals

Experiments were done, using a batch crystallizer apparatus as shown in Fig.1, varying the Mg/P molar ratio from 1-4. The experimental conditions are shown in Table 2. During the process of batch crystallization, the mixed suspension was sampled at the determined crystallization time. The samples were filtered through 0.45 μ m membrane filter, and the concentrations in the filtrate were measured. Also sampled crystals were obtained when the concentration was observed to be constant, to be observed by SEM and be analyzed by the Xray diffractometer. The composition of crystals was analyzed after dissolution by a pH 2 HCl solution.

Results and Discussion

Crystallization characteristics of MAP

Fig.2 shows on outline of the MAP crystallization characteristics in the Mg^{2+}-NH_4^+-PO_4^{3-}-OH^--H_2O system. $MgHPO_4 \cdot 3H_2O$ was produced at higher temperatures. As for the effect of pH, crystals were not produced below pH 7, but $Mg_3(PO_4)_2 \cdot 4H_2O$ and $Mg(OH)_2$ coprecipitated together with MAP at higher pH. Also $Mg(OH)_2$ was produced in the case of excess of magnesium ion. In the case of excess of phosphate ion, another type of crystals was produced.

It was recognized that excess of ammonium ion and suitable pH and Mg/P molar ratio were desirable as the operating conditions for selective MAP crystallization.

Table 1. Experimental conditions

H_3PO_4	[mol/l]	0.0010
NH_4Cl	[mol/l]	0.030
$MgCl_2$	[mol/l]	0.0010
		~0.0020
pH	[-]	8.0~12.0
Agitation rate	[rpm]	300
Operational temperature	[K]	298

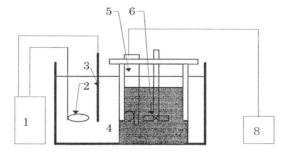

1. Digital controller
2. Heater
3. Thermostat sensor
4. Thermostat bath
5. PH sensor
6. Impeller
7. Crystallizer
8. PH meter

Figure 1. Schematic diagram of experimental apparatus

Table 2. Experimental conditions

H_3PO_4	[mol/l]	0.0010
NH_4Cl	[mol/l]	0.030
$MgCl_2$	[mol/l]	0.0010
		~0.0040
pH	[-]	9.0
Agitation rate	[rpm]	300
Operational temperature	[K]	298

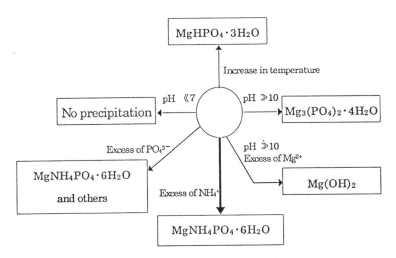

Figure 2. Characteristics of MAP precipitation

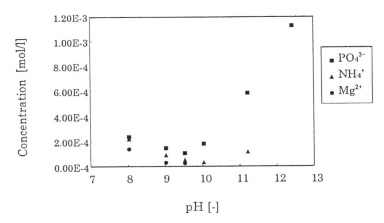

Figure 3. Effect of pH on the equilibrium concentration of each component

Effect of pH on the solubility of MAP

Fig.3 shows the effect of pH on the solubility of MAP. By increasing the pH, the concentration of each reactant is lowered, to reach the lowest concentration in the range of pH from 9 to 10. However the concentration increased when the operational pH became higher than 10. The reason was thought to be that $Mg(OH)_2$ precipitated at the higher pH.

Fig.4 also shows the effect of pH on the phosphate removal efficiency at two different Mg/P ratios. The highest removal efficiency was obtained in the pH range from 9 to 10. Increasing Mg/P molar ratio increased the phosphate removal efficiency.

Effect of Mg/P molar ratio on the behavior of concentration decrease and the properties of produced crystals

SEM photographs of produced crystals are shown in Photos 1,2 and 3, where the Mg/P molar ratio is varying from 1 to 4. Also Xray diffraction analysis of each crystal are shown in Figs. 5,6 and 7. At Mg/P molar ratio 1, the shape of crystals was observed to be one of typical MAP. However the crystals were agglomerated at Mg/P molar ratio 2. When Mg/P molar ratio became 4, the shape of crystals became needlelike and in addition, fine crystals were observed. These crystals were found to be MAP by Xray diffraction analysis, although the shape of the crystals was different. But the results of the composition analysis shown in Table 3, indicate the amount of ammonium ions in the case of Mg/P molar ratio 2 and 4, was higher than in the case of Mg/P molar ratio 1. The reason for the excess uptake of ammonium ion was not clear, but this phenomenon is important when considering the ammonium ion removal.

Figs.8 and 9 show the time course of concentration decrease at different Mg/P molar ratio. At the condition of Mg/P molar ratio 1, the time needed for the concentration to start decreasing was longer, and after that the concentration gradually decreased. But at the condition of Mg/P molar ratio 2 and 3, concentration rapidly decreased from the start of the experiment, to reach the almost constant concentration. These phenomena were thought to come from the increasing nucleation rate with increasing Mg/P molar ratio. This fact is thought to affect the properties of the crystals.

Proposal of a new process for the application of MAP crystallization

Fig.10 shows the application of the concept of MAP crystallization in to sewage treatment process. If MAP crystallization is applied, about 90% of phosphate and 20% of ammonium ion is calculated to be removed and recovered in the form of MAP. The residual ammonium ion is removed by the biological ammonium removal process.

Conclusion

Experiments of reactive crystallization of MAP were performed to obtain the optimal operational conditions for MAP selective crystallization, which were an excess of ammonium ions, pH 9-10, and Mg/P molar ratio 1-4. It was also found that Mg/P molar ratio affected the behavior of the concentration decrease, and changed the properties of the produced crystals. Based on the results of these experiments, MAP crystallization is proposed as an efficient method for the removal of phosphate and ammonium ions from the wastewater of the sludge treatment.

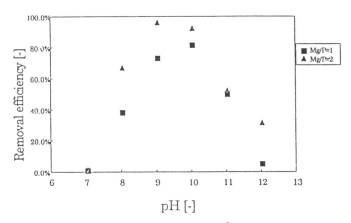

Figure 4. Removal efficiency of PO_4^{3-} depended on pH

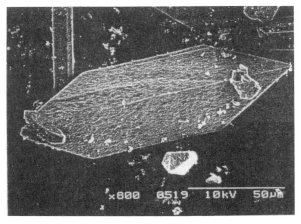

Photo 1. SEM photograph of produced crystals ($Mg^{2+}/PO_4^{3-}=1$)

Photo 2. SEM photograph of produced crystals ($Mg^{2+}/PO_4^{3-}=2$)

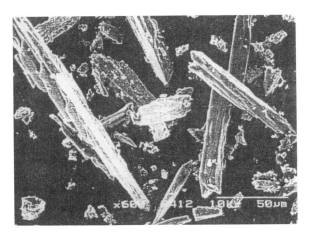

Photo 3. SEM photograph of produced crystals ($Mg^{2+}/PO_4^{3-}=4$)

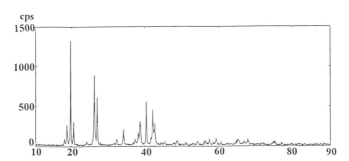

Figure 5. Xray analysis of produced crystals ($Mg^{2+}/PO_4^{3-}=1$)

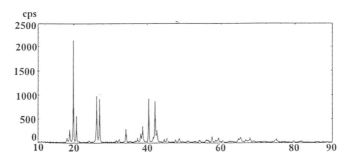

Figure 6. Xray analysis of produced crystals ($Mg^{2+}/PO_4^{3-}=2$)

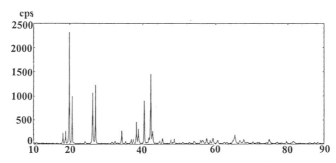

Figure 7. Xray analysis of produced crystals ($Mg^{2+}/PO_4^{3-}=4$)

Table 3. Composition of MAP

$Mg^{2+}/PO4^{3-}$	$PO4^{3-}$	NH_4^+	Mg^{2+}
1	1.00	1.00	1.03
2	1.00	1.25	1.09
4	1.00	1.27	1.06

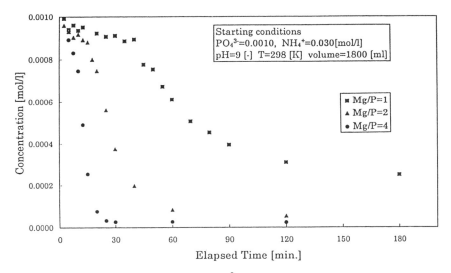

Figure 8. Time course of PO_4^{3-} conc. in solution (1800ml)

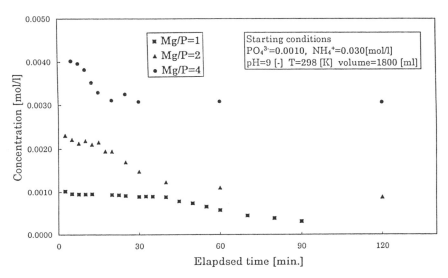

Figure 9. Time course of Mg conc. in solution (1800ml)

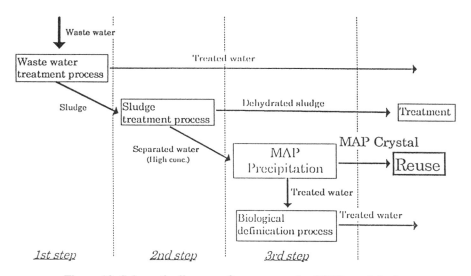

Figure 10. Schematic diagram of a process using MAP precipitation

Literature Cited

1) Hirasawa. I. etal, *Crystallization as a Separation Processes*, **1990**, *ACS Symp. Ser.438 Chapter 26*, pp*353-363*.
2) L.Murrel etal, *Effluent Water Treatment Journal*, **1972,** *Oct*, pp*509-519*.

Chapter 23

Inclusion of Selenium in Ferrite Crystals

Y. Taguchi[1], M. Ohizumi[2], K. Aoyama[3], and K. Katoh[1]

[1]Department of Chemistry and Chemical Engineering, [2]Waste Liquids Treatment Facilities, and [3]Research Institute for Hazards in Snowy Areas, Niigata University, Ikarashi 2-Nocho, Niigata 950–21, Japan

Various heavy metals are included in ferrite crystals when a solution containing the metallic ions is ferritized with Fe(II). To determine the capacity of the ferrite as a scavenger for selenium removal from wastewater, artificially prepared wastewater containing selenium was treated using the ferrite method. The final concentration of selenium in the treated water was less than approximately 0.1 mg/l after a wastewater containing selenium at a concentration of less than 400 mg/l was treated. The amount of selenium removed by ferritization was not influenced by the valence state of selenium, Se(IV) or Se(VI). However, differently distributed selenium in the ferrite particles was found by dissolving the ferrite crystals with an acidic solution. Se(IV) is almost uniformly distributed when the wastewater containing Se(IV) was treated, but when water containing Se(VI) was treated, the higher concentration of Se(VI) was near the center of the ferrite particles.

The element selenium is useful for the manufacturing of chemicals, glasses, ceramics, semiconductors, photocells, photosensitive dyes for copiers, etc., but it is a poisonous substance which is detrimental to human health (1). In water, selenium usually exists in the form of selenite(IV) or selenate(VI), the conjugate acids being selenious acid and selenic acid, which chemically are classified as oxoacids. Thermodynamically, selenate is more stable than selenite. Selenium should be removed from wastewater because of its health hazard at values even lower than, e.g., 0.1 mg/l (2). The removal of selenium is a practical and interesting problem to those who usually apply the ferrite method, a ferritization process, to waste liquids containing heavy metallic ions. Using the process under a given set of operating conditions, some liquids containing selenate or selenite were treated to investigate the mechanism and amount of selenium removed by the ferritization method into the ferrite crystals. Attention was also paid to the effect of ion valency, i.e., four-valent and six-valent, because it was suspected from the inclusion results of arsenic (3) that there could be a difference in the distributions of selenium in the ferrite particles depending on to its valence. This anticipation was experimentally tested.

The aim of this study was to include selenium as an actual impurity in the ferrite crystals under a reaction-crystallization. We let the ferrite crystals play the role of a scavenger as we wanted to produce selenium-free solutions.

Experimental

The experimental procedure is shown in Figure 1. The ferritization reaction (4, 5) was carried out using two artificially prepared model solutions of wastewater containing selenium, Se(IV) or Se(VI). For the preparation of the wastewater, SeO_2 , selenite, tetravalent selenium or K_2SeO_4, selenate, hexavalent selenium was dissolved. A sample solution of 700 ml was poured into a one-liter glass reactor-vessel, and 35 g of $FeSO_4 \cdot 7H_2O$ was added to the solution. The temperature of the solution was then raised to 65 ℃ using an electrical heater with a stirrer. Air was admitted through a distributor immersed in the solution at the rate of 1.0 l/min to start the reaction, and the operation was continued for almost one hour to form the ferrite which included the selenium. The reaction time of one hour is usually necessary to simultaneously remove various heavy metallic ions from the wastewater. During the reaction period, the solution was held at a constant pH of 9 and at a constant temperature of 65 ℃. Using a reflux condenser attached to the reaction-vessel, vapor was condensed and returned to the vessel. This procedure is referred to as the standard ferritization method which is often applied on an industrial plant scale (6). The resultant ferrite crystals were separated from the solution by centrifugation or filtration. Using the separated ferrite crystals, dissolution tests and leaching tests were performed.

Results and Discussion

Residual Concentration of Selenium after Ferritization   In Figure 2, the initial and residual concentrations of selenium during the application of the standard ferritization method are depicted. The initial concentration of selenium was varied from 50 to 500 mg/l. From Figure 2, it can be seen that when the standard ferritization method was applied to the wastewaters containing less than 400 mg/l selenium, the amounts of selenium could be reduced to a concentration level of 0.1 mg/l. This level was our goal since at this level the treated water can be discharged into the public sewage system (2). A large difference in the residual concentrations between Se(IV) and Se(VI) was not found, whether the wastewater contained Se(IV) or Se(VI). When the solution of 500 mg/l was treated, the residual concentration exceeded 1.4 mg/l.

From the observation of residual concentration changes in iron and selenium during ferritization, as shown in Figures 3-1 and 3-2, their minimum values were usually found at almost 40 min after the start of the ferritization reaction. Stopping the reaction at 40

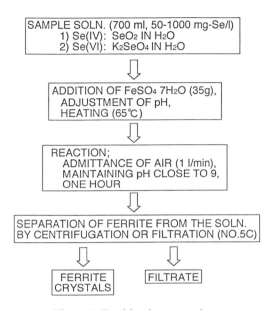

Figure 1. Ferritization procedure.

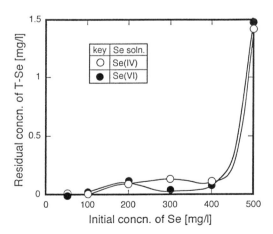

Figure 2. Residual concentration of total selenium
in the treated water after ferritization.

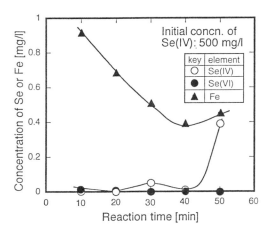

Figure 3-1. Concentration changes of Se(IV), Se(VI) and iron during ferritization of Se(IV) soln.

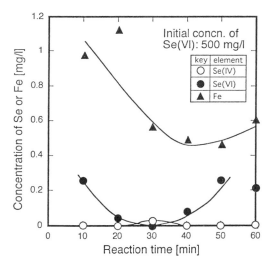

Figure 3-2. Concentration changes of Se(IV), Se(VI) and iron during ferritization of Se(VI) soln.

min enabled the residual concentration of selenium to be less than 0.1 mg/l for the treatment of the 500 mg-Se/l solution.

Distribution of Selenium in Ferrite Crystals   Some dissolution tests of the resultant ferrite crystals were performed to investigate how selenium was distributed in the crystals. About 10 g of ferrite crystals which were usually obtained from each ferritization, were dissolved. They were gradually dissolved with several acidic solutions from the surface to the center. The dissolution operation was repeated six times in order to thoroughly dissolve the residual ferrite crystals. The distributions of selenium in the crystals formed with Se(IV) and Se(VI) were found as shown in Figures 4-1 and 4-2, respectively. The reduced radius, r/R, on the abscissa was determined by the amount of dissolved iron. The ratio of Se/Fe on the vertical axis was the average value calculated from the amounts of iron and selenium dissolved in each acidic solution.

Different distributions of selenium were observed between the crystals formed with selenite and the crystals formed with selenate. From Figure 4-1, almost uniformly distributed Se(IV) is recognized. Near the surface, Se(IV) is slightly higher than that near the center. Near the center, a small amount of Se(VI) was also detected. The partial oxidation of Se(IV) to Se(VI) was considered to be due to the following reaction with oxygen in air.

$$2SeO_3^{2-} + O_2 \rightarrow 2SeO_4^{2-}$$

As ferritization proceeds, a part of the resultant ferrite Fe_3O_4 also acts as a oxidizing agent according to:

$$SeO_3^{2-} + Fe_3O_4 + 3H_2O \rightarrow SeO_4^{2-} + 3Fe(OH)_2$$

These reactions spontaneously occur according to the oxidation-reduction potential in basic solution (7, 8).

From Figure 4-2, Se(VI) was found concentrated around the center. It decreased linearly with the radius, and approached a constant ratio of 0.003 near the surface. At the reduced radius between 0.25 and 0.4, a very small amount of reduced selenium, Se(IV), was detected. The amount was almost less than the detectable limit, but reduction reaction below might occur. The reaction rate of this reduction of Se(VI) was considered to be very slow.

$$SeO_4^{2-} + 2Fe(OH)_2 + H_2O \rightarrow SeO_3^{2-} + 2Fe(OH)_3$$

The smaller amount near the surface is desirable for the inclusion of selenium, because even if the ferrite surface were subject to dissolution, the amount of selenium dissolved would be lower.

In the former crystals, selenite was almost uniformly distributed in the ferrite crystals, but in the latter, selenate was concentrated near the center of the crystals. The reason why Se(VI) was included more deeply or more tightly into the ferrite crystals is not presently understood. However, it is possible to speculate that one of the reasons may be due to its ionic radius and ionic valence. The radius of Se(VI) is smaller than

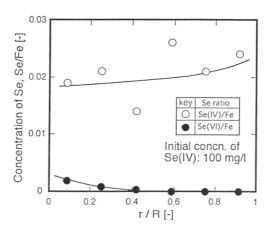

Figure 4-1. Distribution of selenium in the ferrite crystals formed from Se(IV) soln.

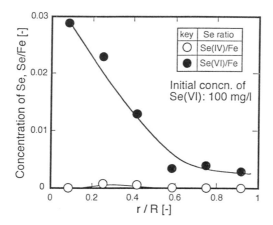

Figure 4-2. Distribution of selenium in the ferrite crystals formed from Se(VI) soln.

that of Se(IV). Smaller ions are capable of remaining in ferrite crystals. However, during ferritization or crystallization, larger species are squeezed towards the liquid side beyond the interface of the ferrite crystal as impurities. This leads to a homogeneous distribution of Se(IV) and a heterogeneous distribution of Se(VI). But, if the concentration of selenium is higher than, e.g., 500 mg/l in the wastewater, the ferritization is not perfect. Ferrite crystals containing selenium at a concentration of 3 wt % could not be obtained.

A similar phenomenon was observed with the inclusion of arsenic (3) when it is also treated by the ferritization method. As(III), having a larger radius, was uniformly distributed in the ferrite crystals formed from As(III) solution. When As(V), having a smaller radius, was treated, a large amount of it was concentrated near the center of the crystals with small amounts of reduced As(III) near the surface.

Stability of the Ferrite Crystals   Leaching tests were carried out to determine the stability of ferrite crystals in water. Using about 10 g of the same ferrite crystals, a leaching test was performed six times with different 100 ml portions of deionized water. The mass ratio of the ferrite crystals to water was 1:10. The flask containing the ferrite crystals and water was shaken for six hours at room temperature at a rate of two hundred times a minute. Figure 5 shows the amount of total selenium leached from the ferrite formed from the Se(IV) or Se(VI) solution using the standard ferritization. If a line of 15 μ g/g-ferrite, equivalent to a concentration of 1.5 mg-Se/l, is supposed to be the tentative limiting value for selenium allowed to be discharged in the treated water, amounts of selenium above the line were observed in the 1st, 2nd, 3rd and 4th leaching tests of the ferrite particles formed from Se(IV) solution. However, the amounts of selenium leached from the ferrite formed from the Se(VI) solution were below the limiting line except for the amount in the 2nd test. For the 5th and 6th leaching tests, selenium was hardly found. Therefore, the ferrite particles containing Se(IV) are unsuitable for subsequent management, e.g., underground reclamation, solidification with concrete. Much attention should be paid to Se(IV) during the ferritization of selenium-containing wastewater.

From the results of the dissolution and leaching tests, the ferrite including selenate was more stable than that including selenite. This suggested that if the wastewater contained selenium in the form of Se(IV), the selenite should be oxidized to selenate before ferritization because a large amount of selenate is incorporated into the center of the ferrite particles. However, the complete oxidation of Se(IV) to Se(VI) in the wastewater is not very easy, and requires the addition of concentrated HCl solutions and heating at nearly 100 ℃ for one hour.

Effects of the Addition of Hydrogen Peroxide and Magnetite-coating   Two attempts were performed to suppress the amount of selenium leached from the ferrite crystals to a lower level than that observed in Figure 5 where the standard ferritization method was used. One was the addition of hydrogen peroxide to the selenium solution, the other was the coating (9) of the surface of the ferrite particles by magnetite. The magnetite was formed from the Fe(II) solution only.

One ml of 35 wt%-H_2O_2 solution was added to a 700 ml solution of Se(IV). After the addition of hydrogen peroxide to the selenite-containing wastewater, the mixture solution was ferritized and the leaching test was performed as shown in Figure 6. The amount of selenium leached was significantly decreased if compared with the values in Figure 5. The amount of selenium leached during the 2nd test appearing in Figure 6 was 0.92 μ g/g-ferrite, which is about 1/60 of the value of 61 μ g in Figure 5. The amount of iron was also decreased from 312 μ g (not shown) to 4.1 μ g during the 2nd leaching test. The addition of a small amount of hydrogen peroxide to the Se(IV) solution is effective against secondary environmental pollution. The excess hydrogen peroxide might share the role of partially oxidizing the Fe(II).

The second attempt was the application of the magnetite-coating method, a process by which the surface of the ferrite particle is thoroughly coated with magnetite, which reduces the amount of leached selenium. The coating operation was repeated 50 times by producing a small degree of supersaturation, ΔC, again and again. It is expected that a slightly soluble substance will grow as a crystal. The results shown in Figure 7 are from when a Se(VI) solution was ferritized. The amount of selenium leached was remarkably decreased if compared to the 2nd test; a decrease of Se from 0.92 μ g in Figure 6 to 0.16 μ g was observed.

Particle Size and Crystallinity of Ferrite Crystals   The particle size of ferrite crystals was measured because the larger particles are, of course, desirable for their easy separation from a suspension. The size was measured using Laser Diffraction Analysis (LDA). Just before the measurement, the suspension containing a small amount of ferrite crystals was sonicated to disperse the ferrite particles. The particle size distribution of the ferrite crystals formed from 400 mg-Se(IV) solution is shown in Figure 8-1. The distribution has two peaks at nearly 0.2 and 1.8 μ m. The median, mode and average diameters were slightly smaller than those obtained from the ferrite crystals formed from only iron(II) sulfate solution.

The mode diameters of the ferrite particles formed from various solutions of selenium are shown in Figure 8-2. The initial concentration of selenium was varied from 0 to 1000 mg/l. The white squares denote the particles formed from the only $FeSO_4$-containing solution without selenium. The mode diameter was gradually decreasing with increasing initial concentration of selenium. The median diameter and average diameter were also decreasing with increasing concentration.

The X-ray diffraction (XRD) intensity of the ferrite crystals formed from various solutions of selenium was observed (Figure 9) in order to examine the crystallinity of ferrite particles. It is preferable to make ferrite particles with high crystallinity. The relative intensity at a d-value of 2.530 (2 θ was about 35.5 degrees, the strongest value among three d-values) was shown with respect to the initial concentration of selenium.

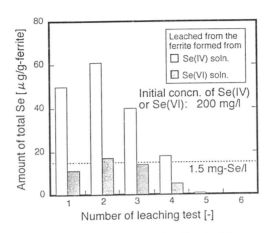

Figure 5. Amount of total selenium leached from the ferrite crystals formed from Se(IV) or Se(VI) soln.

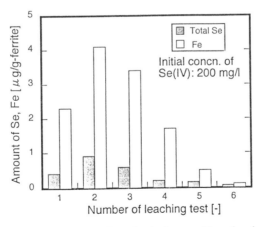

Figure 6. Amounts of total selenium and iron leached from the ferrite crystals formed from Se(IV) soln. to which H_2O_2 soln. was added.

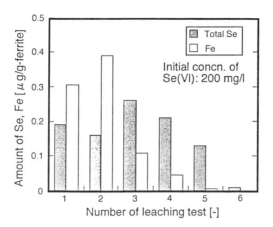

Figure 7. Amounts of selenium and iron leached
from the ferrite crystals coated with magnetite,
following the ferrite formation with Se(VI) soln.

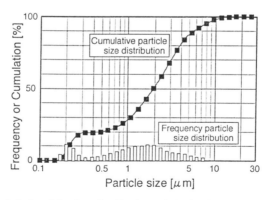

Figure 8-1. Particle size distribution of the ferrite crystals formed
from 400 mg-Se(IV) soln., by SALD-3000 with sonicator.
(Median dia.; 1.930 μm, mode dia.; 2.616, avg. dia.; 1.505)

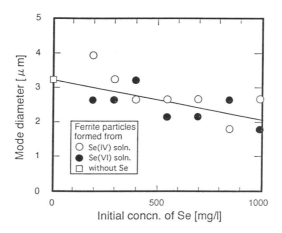

Figure 8-2. Mode diameter of ferrite particles formed from solutions containing Se(IV) or Se(VI).

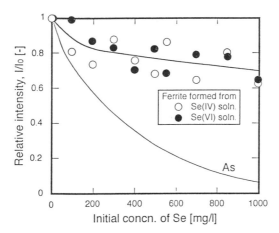

Figure 9. XRD intensity of ferrite crystals formed from the solutions containing Se(IV) or Se(VI).

The intensity without selenium was regarded as unity. The gentle slope shows a gradual decrease in intensity, and means a gradual deterioration of the ferrite crystallinity. For reference, the relative intensity of arsenic-containing ferrite, or the deterioration of the ferrite crystallinity is also shown. It is much steeper.

Conclusions

The capability of ferrite crystals as scavengers for selenium during ferritization accompanied by crystallization was evaluated and the following results were obtained. (1) It is recommended that the concentration of selenium in wastewaters of about 400 mg/l is suppressed before ferritization, when the reaction was performed under normal operating conditions. This easily enables the residual concentration of selenium to be less than about 0.1 mg/l. (2) It is concluded from the leaching tests of ferrite crystals including selenite or selenate, that a relatively larger amount of selenium is leached from the crystals formed from Se(IV) solutions compared to that leached from the crystals formed from Se(VI) solutions. When selenium was leached, a large amount of iron was also leached in proportion to the amount of selenium. Therefore, it is desirable to add a small amount of hydrogen peroxide before ferritization when the wastewater contains Se(IV). (3) From dissolution tests of ferrite crystals with dilute sulfuric acid solution, it was found that selenate was more tightly or more deeply included in the crystals compared to selenite. The dissolution of the crystals formed from Se(VI) solutions proved that a large amount of selenate was concentrated around the center of the crystals. In the ferrite crystals formed from Se(IV) solutions, the selenite was uniformly distributed in the crystals.

Literature Cited

1. National Research Council, *Medical and Biologic Effects of Environmental Pollutants, Selenium*, National Academy of Sciences, Washington, D.C., 1976.
2. Cabinet of Japan, *Kanpo (Japanese Official Daily Gazette)*, Standard of Waste Water, 1993, Dec. 27, No.1308, p.8.
3. Y. Taguchi, M. Ohizumi, K. Katoh and T. Suzuki, *Proc. of 1st Asian Symp. On AAWM*, 1993, p.172.
4. E. W. Gorter, *Philips Res. Rept.*, 1954, No.9, 295.
5. M. Kiyama, *Bull. Chem. Soc. Japan*, 1974, **47**, No.7, 1646.
6. T. Akashi, I. Sugano, Y. Kenmoku, Y. Shinma and T. Tsuji, *NEC Research & Development*, 1970, No.19, 94.
7. K. Morinaga, *Kagaku Sensho No.9, Sanka to Kangen (Oxidation and Reduction)*, Shokabo, Tokyo, 1989, p.111.
8. Chemical Society of Japan, *Kagaku Binran (Chemical Handbook)*, Maruzen, Tokyo, 1984, 3rd ed., p.II-473.
9. Y. Tamaura, P.Q. Tu, S. Rojarayanont and H. Abe, *Wat. Sci. Tech.*, 1991, **23**, 399.

INDEXES

Author Index

Affiliation Index

Subject Index

Bestsellers from ACS Books

The ACS Style Guide: A Manual for Authors and Editors
Edited by Janet S. Dodd
264 pp; clothbound ISBN 0–8412–0917–0; paperback ISBN 0–8412–0943–X

Writing the Laboratory Notebook
By Howard M. Kanare
145 pp; clothbound ISBN 0–8412–0906–5; paperback ISBN 0–8412–0933–2

Career Transitions for Chemists
By Dorothy P. Rodmann, Donald D. Bly, Frederick H. Owens, and Anne-Claire Anderson
240 pp; clothbound ISBN 0–8412–3052–8; paperback ISBN 0–8412–3038–2

Chemical Activities (student and teacher editions)
By Christie L. Borgford and Lee R. Summerlin
330 pp; spiralbound ISBN 0–8412–1417–4; teacher edition, ISBN 0–8412–1416–6

Chemical Demonstrations: A Sourcebook for Teachers, Volumes 1 and 2, Second Edition
Volume 1 by Lee R. Summerlin and James L. Ealy, Jr.
198 pp; spiralbound ISBN 0–8412–1481–6
Volume 2 by Lee R. Summerlin, Christie L. Borgford, and Julie B. Ealy
234 pp; spiralbound ISBN 0–8412–1535–9

From Caveman to Chemist
By Hugh W. Salzberg
300 pp; clothbound ISBN 0–8412–1786–6; paperback ISBN 0–8412–1787–4

The Internet: A Guide for Chemists
Edited by Steven M. Bachrach
360 pp; clothbound ISBN 0–8412–3223–7; paperback ISBN 0–8412–3224–5

Laboratory Waste Management: A Guidebook
ACS Task Force on Laboratory Waste Management
250 pp; clothbound ISBN 0–8412–2735–7; paperback ISBN 0–8412–2849–3

Reagent Chemicals, Eighth Edition
700 pp; clothbound ISBN 0–8412–2502–8

Good Laboratory Practice Standards: Applications for Field and Laboratory Studies
Edited by Willa Y. Garner, Maureen S. Barge, and James P. Ussary
571 pp; clothbound ISBN 0–8412–2192–8

For further information contact:

American Chemical Society
1155 Sixteenth Street, NW ◆ Washington, DC 20036
Telephone 800–227–9919 ◆ 202–776–8100 (outside U.S.)

The ACS Publications Catalog is available on the Internet at
http://pubs.acs.org/books